COOPERATIVE DECISION-MAKING UNDER RISK

THEORY AND DECISION LIBRARY

General Editors: W. Leinfellner (*Vienna*) and G. Eberlein (*Munich*)

Series A: Philosophy and Methodology of the Social Sciences

Series B: Mathematical and Statistical Methods

Series C: Game Theory, Mathematical Programming and Operations Research

SERIES C: GAME THEORY, MATHEMATICAL PROGRAMMING AND OPERATIONS RESEARCH

VOLUME 24

Editor: S. H. Tijs (University of Tilburg); *Editorial Board:* E.E.C. van Damme (Tilburg), H. Keiding (Copenhagen), J.-F. Mertens (Louvain-la-Neuve), H. Moulin (Durham), S. Muto (Tokyo University), T. Parthasarathy (New Delhi), B. Peleg (Jerusalem), H. Peters (Maastricht), T. E. S. Raghavan (Chicago), J. Rosenmüller (Bielefeld), A. Roth (Pittsburgh), D. Schmeidler (Tel-Aviv), R. Selten (Bonn), W. Thomson (Rochester, NY).

Scope: Particular attention is paid in this series to game theory and operations research, their formal aspects and their applications to economic, political and social sciences as well as to socio-biology. It will encourage high standards in the application of game-theoretical methods to individual and social decision making.

The titles published in this series are listed at the end of this volume.

COOPERATIVE DECISION-MAKING UNDER RISK

by

JEROEN SUIJS

Tilburg University

KLUWER ACADEMIC PUBLISHERS

BOSTON / DORDRECHT / LONDON

Distributors for North, Central and South America:
Kluwer Academic Publishers
101 Philip Drive
Assinippi Park
Norwell, Massachusetts 02061 USA
Telephone (781) 871-6600
Fax (781) 871-6528
E-Mail <kluwer@wkap.com>

Distributors for all other countries:
Kluwer Academic Publishers Group
Distribution Centre
Post Office Box 322
3300 AH Dordrecht, THE NETHERLANDS
Telephone 31 78 6392 392
Fax 31 78 6546 474
E-Mail <orderdept@wkap.nl>

Electronic Services <http://www.wkap.nl>

Library of Congress Cataloging-in-Publication Data
A C.I.P. Catalogue record for this book is available
from the Library of Congress. LC# 99047429

Printed on acid-free paper.

Printed in the United States of America

Contents

Acknowledgements

Some four years ago, I started to participate in a research project that was initiated by Stef Tijs at Tilburg University. The aim was to "... develop a theory on games with stochastic payoffs by using traditional cooperative game theory as a guideline." The result is this book on stochastic cooperative games.

Like a true *cooperative* game theorist, I did not do it all on my own. Most of the research has been carried out in close collaboration with Peter Borm, Anja De Waegenaere, and Stef Tijs. I owe them many thanks. Their insights, knowledge, and comments proved to be of inestimable value. In fact, this book would not have existed without them.

I also take this opportunity to thank the many colleagues for their helpful comments and suggestions given at the various seminars and international conferences, where most parts of this book were presented. Finally, I would like to thank Tilburg University, the Netherlands Organization for Scientific Research (NWO), and the Royal Netherlands Academy of Arts and Sciences (KNAW) for their financial support, and the CentER for Economic Research and the Department of econometrics for providing a pleasant and stimulating working environment.

Notations

Many of the notations we use are defined in the text at their first appearance. The following symbols and notations are used throughout this monograph.

The set $\{1, 2, \ldots\}$ of natural numbers is denoted by $\mathbb{N}$ and the set of real numbers is denoted by $\mathbb{R}$. For any finite set N we denote the power set by 2^N, that is, $2^N = \{S | S \subset N\}$, and the number of elements by $\# N$. Furthermore, $\mathbb{R}^N$ denotes the set of all functions from N to $\mathbb{R}$ and a vector $x = (x_i)_{i \in N}$ denotes an element of $\mathbb{R}^N$.

The set of real valued random variables with finite expectation is denoted by $L^1(\mathbb{R})$. For a brief survey on probability theory and stochastic variables we refer to Appendix A.

To improve the readability of this text, we try to keep to the following notations as much as possible. Given a finite set N, the elements of $\mathbb{R}^N$ are denoted by lower case letters and subsets of $\mathbb{R}^N$ are denoted by upper case letters. So, $x \in \mathbb{R}^N$ and $X \subset \mathbb{R}^N$. Similarly, random variables belonging to the set $L^1(\mathbb{R})$ are denoted by boldface upper case letters while sets of random variables are denoted by calligraphic letters. Thus, $\boldsymbol{X} \in L^1(\mathbb{R})$ and $\mathcal{X} \subset L^1(\mathbb{R})$.

1 Introduction

Cooperative behavior generally emerges for the individual benefit of the people and organizations involved. Whether it is an international agreement like the GATT or the local neighborhood association, the main driving force behind cooperation is the participant's believe that it will improve his or her welfare. Although these believed welfare improvements may provide the necessary incentives to explore the possibilities of cooperation, it is not sufficient to establish and maintain cooperation. It is only the beginning of a bargaining process in which the coalition partners have to agree on which actions to take and how to allocate any joint benefits that possibly result from these actions. Any prohibitive objections in this bargaining process may eventually break-up cooperation.

Since its introduction in von Neumann and Morgenstern (1944) , cooperative game theory serves as a mathematical tool to describe and analyze cooperative behavior as mentioned above. The literature, however, mainly focuses on a deterministic setting in which the synergy between potential coalitional partners is known with certainty beforehand. An actual example in this regard is provided by the automobile industry, where some major car manufacturers collectively design new models so as to save on the design costs. Since they know the cost of designing a new car, they also know how much cooperation will save them on design expenditures.

In every day life, however, not everything is certain and many of the decisions that people make are done so without precise knowledge of the consequences. Moreover, the risks that people face as a result of their social and economic activities may affect their cooperative behavior. A typical example in this respect is a joint-venture. Investing in a new project is risky and therefore a company may prefer to share these risks by cooperating in a joint-venture with other companies. A joint-venture is thus arranged before the execution of the project starts when it is still unknown whether it will be a success or a failure. Similarly for insurance deals - which can be interpreted as a cooperation between the insured and the insurance company. A car insurance is concluded for the upcoming year and covers the cost resulting from incidents that might happen during that future period. As opposed to joint ventures and insurance, risk sharing need not be the primary incentive for cooperation. In many other cases, cooperation arises for other reasons and risk is just involved with the actions and decisions of the coalition partners. Small retailers, for instance, organize themselves in a buyers' cooperative to stipulate better prices when purchasing their inventory. Any economic risks, however, are not reduced by such a cooperation.

Cooperative decision-making under risk as described above, is the subject of this book. Based on the model introduced in Suijs, Borm, De Waegenaere and Tijs (1999) , we develop a theory for these so-called stochastic cooperative games by using traditional cooperative game theory as a guideline. In traditional game theory, attention has been and still is focused on two issues, namely, which individuals will eventually cooperate with each other, and how should the resulting benefits be divided? Although both questions may be considered equally important, most of the past research on cooperative game theory has focused on the latter question. In fact, the same thing can be said about this book; it does not focus on the problem of coalition formation. Instead, it presumes that one large coalition will be formed, and focuses on the allocation problem of the corresponding benefits.

The book is organized as follows. Chapter 2 introduces the reader into deterministic cooperative game theory. We formulate a mathematical framework to describe cooperative decision-making problems in general. For analyzing these problems we then turn to cooperative game theory. A cooperative game is usually either of two types: a cooperative game with transferable utility, henceforth referred to as a TU-game, or a cooperative game with non-transferable utility, henceforth referred to as an NTU-game. Which type of cooperative games applies depends on the characteristics of the cooperative decision-making problem that is under consideration. Examples of cooperative games are, of course, TU-games and NTU-games, but also chance-constrained games. In Chapter 2 we provide a brief survey of each of the three types of games.

We start with explaining when a situation can be modeled as an NTU- or TU-game, followed by a short survey of both areas. For TU-games we provide the definitions of some well-known solution concepts like the core, the Shapley value, and the nucleolus. Furthermore, we state the main results in this regard. With respect to the class of NTU-games, we show to what extent the concepts that we discussed for TU-games also apply to NTU-games. It should be noted, however, that both surveys are rather concise. They confine to providing only those concepts that are needed in the forthcoming chapters on stochastic cooperative games.

In a separate section we discuss chance-constrained games. These games are introduced in Charnes and Granot (1973) to encompass situations where the benefits obtained by the agents are random variables. Their attention is also focused on dividing the benefits of the grand coalition. Although the benefits are random, the authors allocate a deterministic amount in two stages. In the first stage, before the realization of the benefits is known, payoffs are promised to the individuals. In the second stage, when the realization is known, the payoffs promised in the first stage are modified if needed. In several papers, Charnes and Granot introduce some allocation rules for the first stage like the prior core, the prior Shapley value, and the prior nucleolus. To modify these so-called prior allocations in the second stage they defined the two-stage nucleolus. We will discuss all these solution concepts and illustrate them with some examples.

Chapter 3 discusses the basic model of stochastic cooperatives games presented in Suijs et al. (1999). Stochastic cooperative games deal with the same kind of problems as chance-constrained games do, albeit in a completely different way. A drawback of the model introduced by Charnes and Granot (1973) is that it does not explicitly take into account the individual's behavior towards risk. The effects of risk averse behavior, for example, are difficult to trace in this model. The model we introduce includes the preferences of the individuals. Any kind of behavior towards risk, from risk loving behavior to risk averse behavior, can be expressed by these preferences. Another major difference is the way in which the benefits are allocated. As opposed to a two-stage allocation, which assigns a deterministic payoff to each agent, an allocation in a stochastic cooperative game assigns a random payoff to each agent. Furthermore, for a two-stage allocation the agents must come to an agreement twice. In the first stage, before the realization of the payoff is known, they have to agree on a prior allocation. In the second stage, when the realization is known, they have to agree on how the prior payoff is modified. For stochastic cooperative games on the other hand, the agents decide on the allocation before the realization is known. As a result, random payoffs are allocated so that no further decisions have to be taken once the realization of the payoff is known.

Besides the definition of the model, Chapter 3 also provides some examples arising from linear production situations, sequencing situations and financial markets. A separate section is devoted to the preferences of the agents. We show which type of preferences allow for risk averse, risk neutral, or risk loving behavior. Furthermore, we define several conditions that will be imposed upon the preferences in future chapters. In particular, we discuss a special class of preferences such that the random benefits can be represented by a deterministic number, also known as the certainty equivalent. For this specific class Suijs and Borm (1999) showed that a stochastic cooperative game can be reduced to a TU-game.

In Chapter 4 we recapitulate the notions of core, superadditivity, and convexity for the class of stochastic cooperative games as defined in Suijs and Borm (1999). In a nutshell, the core consists of those allocations that please each coalition in the sense that this coalition receives at least as much as it can obtain on its own.

For TU-games, there is the well-known result that a core-allocation exists if and only if the game is balanced. For NTU-games, a balancedness condition also exists, except that in this regard it is only a sufficient condition for nonemptiness of the core. For stochastic cooperative games, we formulate such a balancedness condition for a particular subclass.

Superadditivity and convexity tell us something about how the benefits increase with the coalition size. To be more precise, superadditivity implies that two disjoint coalitions are better off by forming one large coalition. Hence, disjoint coalitions have an incentive to merge. Convexity then implies that this incentive increases as the size of the coalitions increases. Thus, for a coalition it is better to merge with a large coalition than with a small one. For TU-games it is known that every convex game is superadditive. Furthermore, a TU-game is convex if and only if all marginal vectors are core-allocations. For explaining a marginal vector, suppose that the grand coalition is formed in a specific order. Thus we begin with a one-person coalition, then an other agents joins this coalition, followed by another one, and so on. The marginal vector with respect to this order then assigns to each agent his marginal contribution, that is, the amount with which the benefits change when he joins a coalition. With respect to NTU-games, again only weaker results hold. For an NTU-game, two definitions of convexity exist. Both imply superadditivity and a nonempty core. Neither of the two, however, implies that all marginal vectors belong to the core. For stochastic cooperative games, we extend the definitions of superadditivity and convexity for TU-games, along the lines of Suijs and Borm (1999). Our extension of convexity is based on the interpretation that a coalition can improve its payoff more by joining a large coalition instead of a small one. We show that convex stochastic cooperative games are superadditive. Furthermore, we show that convexity not only implies that core-allocations exist, but also that all marginal allocations belong to the core of the game. The reverse of this statement, however, is shown to be false: a stochastic cooperative game is not nccessarily convex if all marginal vectors belong to the core. Finally, we show that this new notion of convexity also leads to a new notion of convexity for NTU-games. Since the same result can be derived for NTU-games, that is, all marginal vectors are core-allocations, this new type of convexity thus differs from the existing notions of convexity for NTU-games.

Chapter 5 introduces two nucleoli for stochastic cooperative games. The nucleolus, a solution concept for TU-games, originates from Schmeidler (1969). This solution concept yields an allocation such that the excesses of the coalitions are the lexicographical minimum. The excess describes how dissatisfied a coalition is with the proposed allocation. The larger the excess of a particular allocation, the more a coalition is dissatisfied with this allocation. For Schmeidler's nucleolus the excess is defined as the difference between the payoff a coalition can obtain when cooperating on its own and the payoff received by the proposed allocation. So, when less is allocated to a coalition, the excess of this coalition increases and the other way around.

Since the nucleolus depends mainly on the definition of the excess, we only need to specify the excesses for a stochastic cooperative game in order to define a nucleolus. Unfortunately, this is not that simple. Defining excess functions for stochastic cooperative games appears to be not as straightforward as for TU-games. How should one quantify the difference between the random payoff a coalition can achieve on its own and the random payoff received by the proposed allocation when the behavior towards risk can differ between the members of this coalition? Moreover, the excess of one coalition should be comparable to the excess of another coalition.

For defining the excess for stochastic cooperative games we interpret the excess of Schmeidler's nucleolus in a slightly different way. Bearing the conditions of the core in mind, this excess can be interpreted as follows. Given an allocation of the grand coalition's payoff we distinguish two cases. In the first case, a coalition has an incentive to part company with the grand coalition. Then the excess equals the minimal amount of money a coalition needs on top of what they already get such that this coalition is willing to stay in the grand coalition. In the second case, a coalition has no incentive to leave the grand coalition. Then the excess equals minus the maximal amount of money that can be taken away from this coalition such that this coalition still has no incentive to leave the grand coalition. This interpretation is used to define several excess functions for stochastic cooperative games.

Once the excess is defined, nucleoli are defined in a similar way as for TU-games. We will show that each nucleolus is a well-defined solution concept for stochastic cooperative games, and that it is a subset of the core, whenever the core is nonempty.

Chapter 6 contains an application of stochastic cooperative games to problems of risk sharing and insurance. The insurance of possible personal losses as well as the reinsurance of the portfolios of insurance companies is modeled by means of a stochastic cooperative game. For by cooperating with insurance companies an individual is able to transfer (part of) his future random losses to the insurance companies. Thus, in doing so, he concludes an insurance deal. Similarly for reinsurance, by cooperating with other insurers an insurance company can transfer (parts of) her insurance portfolio to the other insurers.

In this chapter our attention is focused on Pareto optimal allocations of the risks, and on the question which premiums are fair to charge for these risk exchanges.

A Pareto optimal allocation is such that there exists no other allocation which is better for all participants. We show that there is essentially a unique Pareto optimal allocation of risk. It will appear that this Pareto optimal allocation of the risk is independent of the insurance premiums that are paid for these risk exchanges. For determining fair premiums, we look at the core of the insurance game. We show that the zero utility principle for calculating premiums results in a core-allocation.

Chapter 7 analyzes linear production situations with price uncertainty. In a linear production situation, each individual has access to a production technology that he can use to transform his resources into commodities. After production has taken place, he can sell these commodities on a market at fixed prices. Obviously, each individual applies his resources in such a way that the proceeds of production are maximal. However, they can obtain even larger revenues if they cooperate and pool their resources. For instance, there might be a profitable commodity that an agent cannot produce on his own because he lacks some of the necessary resources. By cooperating with agents who can supply these resources, they can produce the good and cash in the corresponding revenues.

Owen (1975) analyzed linear production situations in a completely deterministic setting. The assumption that prices are fixed, is plausible when production takes only little time. But if this is not the case, prices may change during production so that price uncertainty may arise at the time that people have to decide upon the production plan.

An interesting application that fits into this framework is investment funds. In an investment fund, people individually contribute capital for joint investments in financial assets. In this case, capital can be considered as the resource that is needed for the production of assets. Cooperation then offers the opportunity to invest in a more diversified portfolio.

We will show in this chapter that stochastic linear production games have nonempty cores. In particular, we will show for investment funds how the optimal investment portfolio can be determined. In contrast to the results of Izquierdo and Rafels (1996) and Borm, De Waegenaere, Rafels, Suijs, Tijs and Timmer (1999), who consider a similar problem in a deterministic setting, we will show that the allocation rule that yields each investor the same return on investment, does generally not result in a core-allocation.

2 Cooperative Game Theory

This chapter provides the reader with a brief introduction into cooperative game theory, making the usual distinction between games with transferable utility (TU-games) and games with non-transferable utility (NTU-games). We start with defining a mathematical framework to describe cooperative decision making problems in general. Section 2.2 then shows under which conditions such a problem gives rise to an NTU-game and a TU-game, respectively. By means of an example we try to explain when utility is transferable and when not, and what the consequences are for describing the benefits. Subsequently, we state the formal definition of transferable utility. Section 2.3 provides a brief survey of TU-games. We discuss properties like superadditivity and convexity and define the solution concepts core, Shapley value, and nucleolus. Furthermore, we state the main results concerning these concepts. The succeeding section then shows to what extent these results carry over to the class of NTU-games. Note, however, that both surveys are far from complete. They only provide the concepts necessary for understanding the remainder of this monograph. The chapter ends with a summary of chance-constrained games, in which the benefits from cooperation are stochastic variables. This summary is based on Charnes and Granot (1973), (1976), (1977), and Granot (1977) and states the definitions of the so-called prior core, prior Shapley value, prior nucleolus, and the two-stage nucleolus. Furthermore, we compare these definitions with the corresponding ones for TU-games.

2.1 Cooperative Decision-Making Problems

A cooperative decision-making problem describes a situation involving several persons who can obtain benefits by cooperating. The problems they face are who will cooperate with whom, and how will the corresponding benefits be divided. Obviously, in which coalition someone is going to take part, depends on what part of the benefits this coalition has to offer. So, a coalition is only likely to be formed, if all the members of this coalition agree on a specific distribution of the benefits. Finding such an agreement, however, could be troublesome when these members have mutually conflicting interests.

For analyzing such situations, let us formulate a general mathematical framework that captures various cooperative decision-making problems. For starters, let $N \subset \mathbb{N}$ denote the set of individuals. To specify the benefits from cooperation, let Y be some topological space representing the outcome space, whose interpretation depends on the decision-making problem that is under consideration. Moreover, this outcome space Y is such that the benefits generated by each coalition $S \subset N$, can be represented by a subset $Y_S \subset \prod_{i \in S} Y$. An outcome $(y_i)_{i \in S} \in Y_S$ then yields the payoff y_i to agent i. In order to evaluate different outcomes, each individual $i \in N$ has a preference relation $\succsim_i$ over the outcome space Y. So, agent i is only interested in what he himself receives and he does not take into account the payoffs of the other agents. Then given any two outcomes $y, \hat{y} \in Y$, we write $y \succsim_i \hat{y}$ if individual i finds the outcome y as least as good as the outcome $\hat{y}$. A preference relation $\succsim_i$ is composed of a symmetric part $\sim_i$ and an asymmetric part $\succ_i$. $y \sim_i \hat{y}$ means that agent i is indifferent between the outcomes y and $\hat{y}$, i.e. $y \succsim_i \hat{y}$ and $\hat{y} \succsim_i y$. Similarly, $y \succ_i \hat{y}$ means that agent i strictly prefers outcome y to $\hat{y}$, i.e. $y \succsim_i \hat{y}$ but not $\hat{y} \succsim_i y$. Summarizing, a cooperative decision-making model can be described by a tuple $(N, \{Y_S\}_{S \subset N}, \{\succsim_i\}_{i \in N})$.

Example 2.1 In an exchange economy, each agent is endowed with a certain bundle of consumption goods. Each agent can consume his own bundle, but it could well be possible that an agent possesses some commodities he does not like which he would rather like to exchange these ones for goods he likes better. So, agents may benefit from exchanging their initial endowments.

To formulate this situation as a cooperative decision-making problem, let N represent the individuals participating in the exchange economy, and let $M \subset \mathbb{N}$ denote the set of consumption goods. ω_j^i is the amount that agent $i \in N$ possesses of commodity $j \in M$ and $\omega^i \in \mathbb{R}_+^M$ is agent i's commodity bundle. Since an outcome describes a redistribution of the available consumption goods, the outcome space Y equals the commodity space $\mathbb{R}_+^M$. Next, consider a coalition $S \subset N$ of agents. The total number of commodities available to coalition S equals $\sum_{i \in S} \omega^i$. Hence, the outcome space Y_S equals

$$Y_S = \{(c^i)_{i \in S} \in (\mathbb{R}_+^m)^S | \sum_{i \in S} c^i \leq \sum_{i \in S} \omega^i\},$$

with $c^i \in \mathbb{R}_+^m$ representing the bundle of consumption goods assigned to agent $i \in S$. Note that the $\leq$-sign allows the agent to dispose some of the goods if they want to. A coalition S can benefit from cooperation if there exists an outcome $(c^i)_{i \in S} \in Y_S$ such that $c^i \succ_i \omega^i$ for each $i \in S$, that is, each agent $i \in S$ strictly prefers the bundle c^i to his initial endowment ω^i.

Example 2.2 Consider a group of agents, each having a bundle of resources which can be used to produce a number of consumption goods. For the production of these goods, every agent has access to the same linear production technology. Once the goods are produced, they are sold for given prices. In this situation the agents could individually use their own resources to produce a bundle of consumption goods that maximizes the revenues. But they can probably do better if they cooperate with each other and combine their resources. For instance, there may be a consumption good that an agent cannot produce on his own, because he lacks some of the resources to produce it. Cooperating with agents who do possess the absent resources, then enables them to produce this good, and, subsequently, pocket the corresponding revenues.

This so-called linear production situation can be formulated as a cooperative decision-making problem in the following way. Let $N \subset \mathbb{N}$ denote the set of agents, $R \subset \mathbb{N}$ the set of resources, and $M \subset \mathbb{N}$ the set of consumption goods. Furthermore, let p_j be the price at which good $j \in M$ can be sold. Since the benefits consist of a division of the revenues generated by the sale of consumption goods, an outcome specifies for each individual i his share in the revenues. Hence, the outcome space equals $Y = \mathbb{R}$. The resources that are needed to produce a consumption bundle $c \in \mathbb{R}^M_+$ equal Ac, where $A \in \mathbb{R}^{R \times M}$ represents the linear production technology, which is available to each agent. Now, if $b^i \in \mathbb{R}^R_+$ denotes the resource bundle of agent i, then the consumption bundles c that agent i can produce are $C(\{i\}) = \{c \in \mathbb{R}^m_+ |\ Ac \leq b^i\}$. Hence, the corresponding revenues he can obtain equal

$$Y_{\{i\}} = \{x \in \mathbb{R} |\ \exists_{c \in C(\{i\})} : x_i \leq p^\top c\}.$$

Since a coalition $S \subset N$ can use the resources of all its members, the feasible consumption bundles are $C(S) = \{c \in \mathbb{R}^M_+ |\ Ac \leq \sum_{i \in S} b^i\}$. An outcome for coalition S then equals a distribution of the revenues this coalition can obtain, that is,

$$Y_S = \{(x_i)_{i \in S} \in \mathbb{R}^S |\ \exists_{c \in C(S)} : \sum_{i \in S} x_i \leq p^\top c\}.$$

Finally, it is most likely that the preferences of an agent are such that the more money he receives the better. So, for $x, \hat{x} \in \mathbb{R}$ we have that $x \succsim_i \hat{x}$ if $x \geq \hat{x}$.

Example 2.3 Consider a large firm that consists of several divisions, which all have to make use of the same facility - for instance, a repair and maintenance facility - that can only serve one division at a time. When divisions ask for service of the repair and maintenance facility, each division incurs costs for the time it is waiting for service. Now, suppose that a fixed number of divisions has requested service from this facility and that initially they will be served on a first come first served basis. Since the service demanded by the several divisions need not be equally urgent, a division with relatively high urgency may significantly decrease its waiting costs if it could move further forward in the queue. So, by rearranging their positions in the queue, the divisions might decrease the total waiting costs.

When modeling this sequencing situation as a cooperative decision-making problem, the set $N \subset \mathbb{N}$ denotes the divisions of the firm. Next, denote the order in which the divisions are served by $\sigma : N \to \{1, 2, \ldots, n\}$, with $\sigma(i) = j$ meaning that division i takes position j in the queue. Furthermore, denote the initial serving order based on the 'first come first served' principle by σ_0. Now, it would be straightforward to take the outcome space Y to be the set of all possible serving orders σ. For this situation, however, this outcome space is too restrictive. For if two divisions i and j trade places in the serving order, the waiting time increases for division i, say, and the waiting time decreases for division j. Since an increase in waiting time for division i increases its costs, division i only agrees to trade places with division j, if division i is compensated for its increased waiting costs. To enable such compensations we allow transfer payments of money between the divisions. The outcome space thus becomes $Y = \mathbb{R} \times \Pi_N$, where $\mathbb{R}$ represents the amount of money a division receives and $\Pi_N = \{\sigma : N \to \{1, 2, \ldots, n\} | \forall_{i,j \in N: i \neq j} : \sigma(i) \neq \sigma(j)\}$ denotes the set of all possible serving orders. To determine the outcome space of a coalition S of divisions, note that a restraint is put on the rearrangements this coalition can make with respect to the serving order. Two divisions i and j of a coalition are only allowed to change places if all divisions that are waiting in between these two divisions also participate in the coalition. The outcome space for coalition S thus equals

$$Y_S = \{(x_i, \sigma)_{i \in S} \in \prod_{i \in S} (\mathbb{R} \times \Pi_S) | \sum_{i \in S} x_i \leq 0\},$$

with $\Pi_S = \{\sigma \in \Pi_N | \forall_{i \in S} \forall_{j \notin S} : \sigma(j) < \sigma(i) \Leftrightarrow \sigma_0(j) < \sigma_0(i)\}$ the set of feasible rearrangements for coalition S. The vector $(x_i)_{i \in S}$ represents the transfer payments between the divisions in coalition S. Since the amounts of money received by some divisions have to be paid by the other divisions, we have that $\sum_{i \in S} x_i \leq 0$.

Over the years, several game theoretical models have been introduced to deal with cooperative decision-making problems like the ones presented in Examples 2.1 - 2.3. From all of these models, TU-games and NTU-games certainly are the most commonly used. Other, less frequently appearing cooperative games are, amongst others, multi-criteria games, multi-choice games and chance constrained games. Which of these models best suits a specific cooperative decision-making problem depends on the characteristics of the problem under consideration.

The exchange economy presented in Example 2.1, for instance, can be modeled as an NTU-game if the preferences of each agent can be represented by a utility function, that is for each agent $i \in N$ there exists a function $U_i : \mathbb{R}_+^M \to \mathbb{R}$ such that for any $c^i, \hat{c}^i \in \mathbb{R}_+^M$ it holds that $c^i \succsim_i \hat{c}^i$ if and only if $U_i(c^i) \geq U_i(\hat{c}^i)$, i.e., a bundle c^i is preferred to $\hat{c}^i$ if and only if the utility that agent i gets from c^i exceeds the utility that he gets from $\hat{c}^i$. If, however, the preferences are such that agent i prefers the bundle c^i to $\hat{c}^i$ if and only if $c^i_j \geq \hat{c}^i_j$ for $j = 1, 2, \ldots, m$, then a multi-criteria game applies best. Multi-choice games on the other hand, apply when an agent not only has to decide which coalition to join, but also has to decide upon the activity level of his cooperation. Such situations occur, for example, in construction work, where several firms work together on the construction of a building. Now, if these firms get paid dependent on the completion date, the revenues not only depend on which firms participate in the construction, but also on the effort they put into it.

Multi-criteria games and multi-choice games though, are not further discussed in this monograph. The interested reader is referred to Bergstresser and Yu (1977) and Anand, Shashishekhar, Ghose and Prasad (1995) for multi-criteria games and to Hsiao and Raghavan (1993) and van den Nouweland, Potters, Tijs and Zarzuelo (1995) for multi-choice games. In the forthcoming sections we focus on TU-games, NTU-games, and chance-constrained games only. The next section explains when a cooperative decision-making problem can be modeled as an NTU-game and TU-game, respectively.

2.2 Transferable and Non-transferable Utility

The terms transferable and non-transferable utility refer to the agents' preferences in a cooperative decision-making problem. First of all, they imply that the preferences can be represented by a utility function. Given a cooperative decision-making problem $(N, \{Y_S\}_{S \subset N}, \{\succsim_i\}_{i \in N})$, this means that for each agent $i \in N$ there exists a utility function $U_i : Y \to \mathbb{R}$ such that for any $y, \hat{y} \in Y$ it holds that $y \succsim_i \hat{y}$ if and only if $U_i(y) \geq U_i(\hat{y})$. Not every preference relation, however, can be represented by such a utility function. For this to be the case, consider the following three properties.

(P1) A preference relation $\succsim_i$ is called *complete* if for any two outcomes $y, \hat{y} \in Y$ it holds that $y \succsim_i \hat{y}$, or $\hat{y} \succsim_i y$, or both.

(P2) A preference relation $\succsim_i$ is called *transitive* if for any $y, \hat{y}, \tilde{y} \in Y$ such that $y \succsim_i \hat{y}$ and $\hat{y} \succsim_i \tilde{y}$, it also holds true that $y \succsim_i \tilde{y}$.

(P3) A preference relation $\succsim_i$ is called *continuous* if for every $y \in Y$ the sets $\{\hat{y} \in Y \mid \hat{y} \succ_i y\}$ and $\{\hat{y} \in Y \mid y \succ_i \hat{y}\}$ are open in Y.

Property (P1) states that an agent can rank any two outcomes y and $\hat{y}$. So the case that they are incomparable does not occur. Property (P2) states that if an agent prefers the outcome y to the outcome $\hat{y}$, and he prefers $\hat{y}$ to $\tilde{y}$, then he also prefers y to $\tilde{y}$. Property (P3) is a kind of smoothness condition. Roughly speaking, it states that if an agent strictly prefers the outcome $\hat{y}$ to the outcome y, then he also strictly prefers the outcome $\hat{y}$ to all outcomes $\tilde{y}$ that differ only slightly from the outcome y.

The following result, which can be found in Debreu (1959), provides sufficient conditions on $\succsim_i$ so that it can be represented by a utility function.

Theorem 2.4 Let $\succsim_i$ be a preference relation on a connected topological space Y. If $\succsim_i$ satisfies conditions (P1) - (P3), then there exists a continuous utility function $U_i : Y \to \mathbb{R}$ such that for any $y, \hat{y} \in Y$ we have that $y \succsim_i \hat{y}$ if and only if $U_i(y) \geq U_i(\hat{y})$.

Note that this utility function U_i is not uniquely determined. For if U_i represents the preferences $\succsim_i$, then so does any monotonic transformation of U_i. This means that if $f : \mathbb{R} \to \mathbb{R}$ is a strictly increasing function, then the utility function $\hat{U}_i$ defined by $\hat{U}_i(t) = f(U_i(t))$ for all $t \in \mathbb{R}$ also represents the preference relation $\succsim_i$.

For the remainder of this section assume that the conditions of Theorem 2.4 are satisfied. So, given a coalition S and its outcome space Y_S, we can determine the individual utility levels that the agents in this coalition can obtain. Then a cooperative decision-making model $(N, \{Y_S\}_{S \subset N}, \{\succsim_i\}_{i \in N})$ can be modeled as an NTU-game (N, V), where N is the set of agents and $V(S) \subset \mathbb{R}^S$ is the set of utility levels coalition S can obtain, that is,

$$V(S) = \{(z_i)_{i \in S} \in \mathbb{R}^S \mid \exists_{(y_i)_{i \in S} \in Y_S} \forall_{i \in S} : z_i \leq U_i(y_i)\}. \tag{2.1}$$

Note that for mathematical convenience the set $V(S)$ not only includes the utility levels $(U_i(y_i))_{i \in S}$ for all $(y_i)_{i \in S} \in Y_S$, but also the utility levels $z \in \mathbb{R}^S$ which are worse. Note that the preferences $\succsim_i$ need not be included in the description (N, V) of an NTU-game, because for each agent it holds that the higher his utility the better.

NTU-games were introduced in Aumann and Peleg (1960) and generalize TU-games, which were already introduced in von Neumann and Morgenstern (1944). Whether or not an NTU-game can be reduced to a TU-game, depends on the agents' utility functions. In fact, it depends on whether or not adding the individual utilities makes any sense. If so, the benefits from cooperation may be determined by adding the individual utilities and maximizing the resulting sum over all available outcomes. A TU-game can thus be described by a pair (N, v), where N is the set of agents and

$$v(S) = \max \left\{ \sum_{i \in S} U_i(y_i) \middle| (y_i)_{i \in S} \in Y_S \right\} \tag{2.2}$$

are the benefits for coalition $S \subset N$. Note that, by the same argument as for NTU-games, we do not need to include the preference relations $\succsim_i$, $i = 1, , 2, \ldots, n$, in the description (N, v) of a TU-game.

To explain the difference between transferable and non-transferable utility, consider the following example, in which two agents have to divide an indivisible object among the both of them.

Let us begin with specifying the outcome space Y and $Y_{\{1,2\}}$, respectively. To describe a distribution of the benefits, let the vector $y = (y_1, y_2)$ denote a distribution of the object. Since the object is indivisible, it can be given to either agent 1, to agent 2, or, in case of disposal, to neither of them. Hence, the only possible allocations are $(1, 0)$, $(0, 1)$, and $(0, 0)$. Now, if the object is given to agent 1, then agent 2 receives nothing. Obviously, agent 2 disapproves of this allocation. Similarly, if agent 2 receives the object, then agent 1 gets nothing and he disapproves of the allocation. In order to make some progress in this bargaining process, let us introduce transfer payments. So if, for instance, agent 1 receives the object then he can pay money to agent 2 to compensate him for not receiving the object. Let the vector (x_1, x_2) denote these transfer payments. So, an outcome for each agent states the amount of money he receives and who receives the object. This means that we can take the outcome space Y equal to

$$Y = \mathbb{R} \times \{(1, 0), (0, 1), (0, 0)\}.$$

Now, let us determine the outcome space $Y_{\{1,2\}} \subset Y \times Y$. Since the amount of money received by one person has to be paid by the other one, we have that $x_1 + x_2 \leq 0$. The outcome space $Y_{\{1,2\}}$ thus equals

$$Y_{\{1,2\}} = \{((x_1, y), (x_2, y)) \in Y \times Y | \ x_1 + x_2 \leq 0 \text{ and } y \in \{(1, 0), (0, 1), (0, 0)\}\}.$$

Next, let the preferences of the agents be described by the following utility functions:

$$U_1(x,y) = \begin{cases} x + 200, & \text{if } y = (1,0), \\ x, & \text{if } y \neq (1,0) \end{cases}$$
$$U_2(x,y) = \begin{cases} x + 400, & \text{if } y = (0,1), \\ x, & \text{if } y \neq (0,1), \end{cases}$$

This means that agent i prefers an allocation (x,y) to the allocation $(\hat{x},\hat{y})$ if $U_i(x,y) > U_i(\hat{x},\hat{y})$. For instance, if $(x_1,x_2) = (-100,100)$ and $y = (1,0)$ then agent 1's utility equals $-100+200 = 100$ and agent 2's utility equals 100. Furthermore, if the object is given to agent 2 instead of agent 1, that is, $y = (0,1)$, then agent 1's utility decreases to -100 and agent 2's utility increases to 500. Hence, agent 1 prefers the first allocation to the latter, while agent 2 prefers the latter to the first.

Now, let us start with describing the benefits in terms of NTU-games. For this purpose we determine the utility levels of all possible allocations. First, consider the allocations that assign the object to neither agents, that is, $y = (0,0)$. Then given an allocation (x_1,x_2), agent i's utility equals x_i. Since $x_1 + x_2 \leq 0$, the utility levels they can obtain are

$$\{(x_1,x_2)|\ x_1 + x_2 \leq 0\}.$$

Second, consider the allocations that assign the object to agent 1, that is $y = (1,0)$. Then agent 1's utility equals $x_1 + 200$ and agent 2's utility equals x_2. The utility levels they can obtain in this way are

$$\{(x_1 + 200, x_2)|\ x_1 + x_2 \leq 0\}.$$

Finally, consider the allocations that assign the object to agent 2, that is, $y = (0,1)$. In that case, agent 1's utility equals x_1 and agent 2's utility equals $x_2 + 400$. Hence, the corresponding utility levels are

$$\{(x_1, x_2 + 400)|\ x_1 + x_2 \leq 0\}.$$

As Figure 2.1 shows, the set $\{(x_1, x_2 + 400)|\ x_1 + x_2 \leq 0\}$ contains both the set $\{(x_1 + 200, x_2)|\ x_1 + x_2 \leq 0\}$ and the set $\{(x_1,x_2)|\ x_1 + x_2 \leq 0\}$. Hence, the benefits that both agents can obtain from dividing the object can be described by

$$V(\{1,2\}) = \{(x_1, x_2 + 400)|\ x_1 + x_2 \leq 0\}.$$

Formulating the benefits in this way is typical for NTU-games.

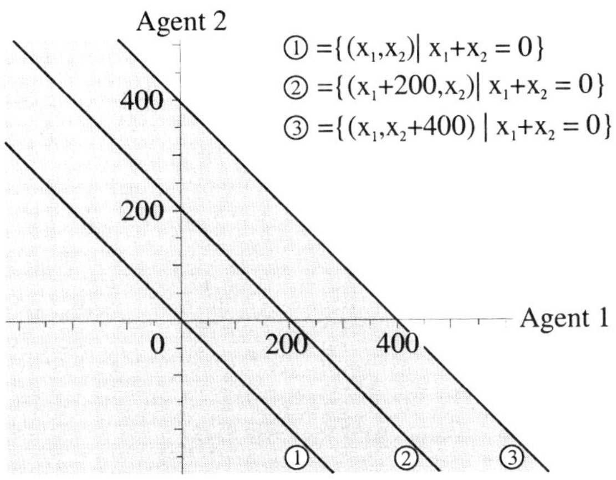

Figure 2.1

Next, let us describe the benefits in terms of TU-games. This means that we add the individual utilities and maximize this sum over all possible allocations. Solving

$$\begin{aligned} \text{maximize} \quad & x_1 + x_2 + (200, 400)y^\top \\ \text{subject to:} \quad & x_1 + x_2 \leq 0, \\ & y \in \{(1,0),(0,1),(0,0)\} \end{aligned}$$

yields a maximum of 400. So, instead of stating that the agents have to divide the object, we can say that they have to divide 400 units of utility. What is left to check is if adding the individual utilities is a justified operation, that is, does it make any difference whether we state the benefits as 400 or as $\{(x_1, x_2+400)|\ x_1+x_2 \leq 0\}$. In answering this question, let us examine the utility levels that both agents can obtain when they may allocate 400 units of utility. If u_i denotes the amount of utility that agent i receives, then the utility levels they can realize are described by the set

$$\{(u_1, u_2)|\ u_1 + u_2 \leq 400\}.$$

This set is pictured in Figure 2.2. Comparing Figure 2.1 to Figure 2.2 reveals that the set $\{(u_1, u_2)|\ u_1 + u_2 \leq 400\}$ coincides with the set $\{(x_1, x_2 + 400)|\ x_1 + x_2 \leq 0\}$. Hence, each allocation $((x_1, y),(x_2, y))$ corresponds with an allocation $u = (u_1, u_2)$ of 400 units of utility, and the other way around; for each allocation $u = (u_1, u_2)$ there exists an allocation $((x_1, y),(x_2, y)) \in Y_{\{1,2\}}$ such that the corresponding utilities coincide, that is, $U_1(x_1, y) = u_1$ and $U_2(x_2, y) = u_2$. This implies that it is indeed correct to state that the agents can divide 400 units of utility.

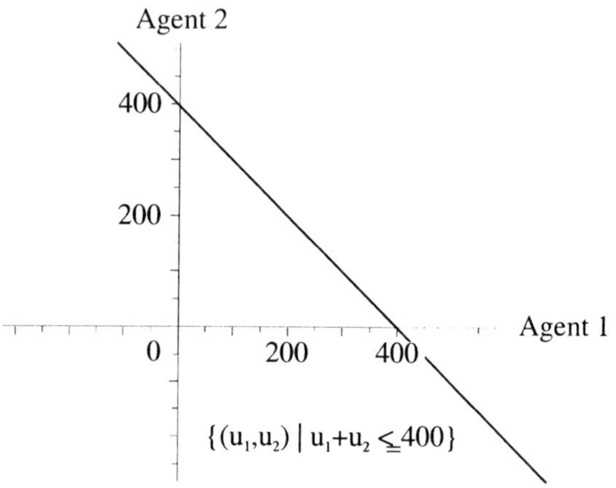

Figure 2.2

Determining the benefits by maximizing the sum of the individual utilities does not work in general. Suppose, for instance, that the utility functions are given by

$$W_1(x,y) = \begin{cases} f(x)+200, & \text{if } y=(1,0), \\ f(x), & \text{if } y \neq (1,0) \end{cases}$$
$$W_2(x,y) = \begin{cases} f(x)+400, & \text{if } y=(0,1), \\ f(x), & \text{if } y \neq (0,1), \end{cases}$$

where

$$f(t) = \begin{cases} 2t, & \text{if } t \leq 0 \\ \frac{1}{2}t, & \text{if } t > 0. \end{cases}$$

So, agent i's utility for money depends on whether he receives money or has to pay it.

Again, let us start with describing the benefits in terms of NTU-games. Then the set of feasible utility levels equals

$$V(\{1,2\}) = \{(f(x_1), f(x_2)+400) | \, x_1+x_2 \leq 0\} \cup \{(f(x_1)+200, f(x_2)) | \, x_1+x_2 \leq 0\}$$

In contrast to the previous situation, describing the benefits by the maximal sum of the individual utilities is not possible. This can be seen as follows. The maximum equals 400 and is attained in the allocation $((x_1,y),(x_2,y)) = ((0,(0,1)),(0,(0,1)),$ which corresponds to the point $(0,400)$ in Figure 2.3. Now, if we state that the benefits equal 400 units of utility, then the agents may allocate this 400 between the two of them. This means that they can obtain the following utility levels

$$\{(w_1,w_2) | \, w_1+w_2 \leq 400\},$$

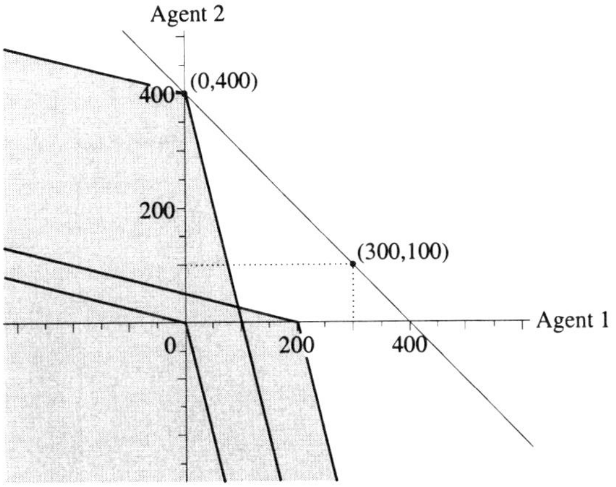

Figure 2.3

which, among others, also includes the allocation $(300, 100)$. But as Figure 2.3 shows, there exists no allocation $((x_1, y), (x_2, y)) \in Y_{\{1,2\}}$ that yields agent 1 and 2 a utility level of 300 and 100, respectively. Apparently, allocating 400 units of utility is not similar to allocating the object. So, what makes the difference then? Why does adding the utility functions U_1 and U_2 not raise any problems, while adding W_1 and W_2 does?

The answer to this question is the following property, which U_1 and U_2 possess, namely, that a transfer of money from one person to another, corresponds with a transfer of utility, hence the distinction between games with transferable utility and games with non-transferable utility. To explain this transferable utility property in more detail, consider an arbitrary allocation $((x_1, y), (x_2, y)) \in Y_{\{1,2\}}$ such that $y = (0, 1)$, that is, agent 2 receives the object. The utility levels $U_1(x_1, y)$ and $U_2(x_2, y)$ that correspond with this allocation thus are x_1 for agent 1 and $x_2 + 400$ for agent 2. Now, suppose that an amount of t dollar is transferred from agent 2 to agent 1, then agent 1 receives t dollar and agent 2 receives the object and $-t$ dollar. This transfer payment increases agent 1's utility with t units to $x_1 + t$, and it decreases agent 2's utility with t units to $x_2 - t + 400$. In other words, transferring t dollar from agent 2 to agent 1 corresponds with transferring t units of utility from agent 2 to agent 1. So instead of transferring money they transfer units of utilities.

A consequence of this transferable utility property is that each agent's valuation for the object does not depend on the amount of money he receives. Indeed, agent 1's utility equals $x_1 + 200$ if he gets the object and x_1 if he does not get it. The difference always amounts to 200, whatever the value of x_1. So agent 1 values the object at 200 units of utility. Similarly, it follows that agent 2 values the object at

400 units of utility.

Since utility is transferable, the more units of utility the agents can divide, the better it is for the both of them. This means that we need to focus on the allocations that maximize the sum of the utilities. This sum equals 0 if no one receives the object, 200 if agent 1 receives the object, and 400 if agent 2 receives the object. Hence, the object should be given the agent 2, so that the benefits equal 400 units of utility.

That the utility functions W_1 and W_2 do not satisfy the transferable utility property is easily seen as follows. Consider the allocation $((x_1, y), (x_2, y)) = ((0, (0, 1)), (0, (0, 1)))$, which corresponds to the point $(0, 400)$ in Figure 2.2. Now, if we transfer \$ 200 from agent 2 to agent 1, so that agent 1 receives \$ 200 and agent 2 pays \$ 200, then agent 1's utility only increases with 100 to $f(200) = 100$. Agent 2's utility, on the other hand, decreases with 400 to $f(-200) + 400 = 0$. Hence, transferring \$ 200 dollar from agent 2 to agent 1 does not transfer 200 units of utility. In fact, it decreases the total utility with 300. Although total utility has decreased, this does not mean that the allocation $((200, (0, 1)), (-200, (0, 1)))$ is worse than $((0, (0, 1), (0, (0, 1)))$; both allocations are optimal in the sense, that there exists no other allocation that yields a higher utility for both agents (see also Figure 2.3).

The example above shows that not every utility function satisfies the transferable utility property. This naturally raises the question, what utility functions do satisfy this transferable utility property, and, maybe even more important, which preferences can be represented by such utility functions? To start with the first question, recall that utility is transferable if a transfer of money between agents corresponds to a transfer of utility. For this to occur in the first place, the outcome space Y must at least allow for transfer payments between the agents. For if agents cannot transfer any money, a transfer of money can certainly not correspond to a transfer of utility. This means that Y can be written as $\mathbb{R} \times \tilde{Y}$, where $\mathbb{R}$ represents the amount of money an agent receives and $\tilde{Y}$ represents all other outcomes (see also Example 2.3 on sequencing situations). Then the outcome space Y_S for coalition S is a subset of $\prod_{i \in S}(\mathbb{R} \times \tilde{Y})$ and the utility that agent $i \in S$ assigns to an outcome $((x_i, y_i))_{i \in S} \in Y_S$ equals $U_i(x_i, y_i)$. Moreover, it satisfies the transferable utility property if the utility function U_i is linearly separable in money, that is, if there exists a function $v_i : \tilde{Y} \to \mathbb{R}$ such that for each outcome $(x_i, y_i) \in \mathbb{R} \times \tilde{Y}$ it holds that

$$U_i(x_i, y_i) = x_i + v_i(y_i). \tag{2.3}$$

To see that utility is indeed transferable, consider two agents i and j such that $U_i(x_i, y_i) = x_i + v_i(y_i)$ and $U_j(x_j, y_j) = x_j + v_j(y_j)$. Let $i, j \in S$ and take an outcome $((x_k, y_k))_{k \in S} \in Y_S$. Next, let τ be a transfer payment of t dollars from agent i to agent j. Hence, agent i receives $x_i - t$ dollar and agent j receives $x_j + t$ dollar. Then $U_i(x_i - t, y_i) = x_i - t + v_i(y_i)$ and $U_j(x_j + t, y_j) = x_j + t + v_i(y_j)$. Since the transfer payment of t dollars from agent i to agent j decreases agent i's utility with t, and increases agent j's utility with t, it corresponds to a transfer of t units of utility. Furthermore, if all utility functions are of the form (2.3) then the benefits of coalition S can be described by the maximum sum of the individual utilities. The proof is straightforward. Since transfer payments between the agents are allowed, this means that for each outcome $((x_i, y_i))_{i \in S} \in Y_S$ it holds that $\sum_{i \in S} x_i \leq 0$. Then given the utility functions $U_i(x_i, y_i) = x_i + v_i(y_i)$ for all $i \in S$ the benefits in terms of an NTU-game equals

$$V(S) = \{z \in \mathbb{R}^S \mid \exists_{((x_i,y_i))_{i \in S} \in Y_S} \forall_{i \in S} : z_i \leq x_i + v_i(y_i)\}.$$

Assuming that the maximum of the sum of the individual utility levels exists, it equals

$$\begin{aligned} v(S) &= \max\left\{\sum_{i \in S}(x_i + v_i(y_i)) \mid ((x_i, y_i))_{i \in S} \in Y_S\right\} \\ &= \max\left\{\sum_{i \in S} v_i(y_i) \mid ((x_i, y_i))_{i \in S} \in Y_S\right\}. \end{aligned}$$

When coalition S allocates $v(S)$, the utility that levels the members of S can obtain equal $\{z \in \mathbb{R}^S \mid \sum_{i \in S} z_i \leq v(S)\}$. Hence, it is sufficient to show that $V(S) = \{z \in \mathbb{R}^S \mid \sum_{i \in S} z_i \leq v(S)\}$. That $V(S) \subset \{z \in \mathbb{R}^S \mid \sum_{i \in S} z_i \leq v(S)\}$ follows immediately from

$$v(S) = \max\{\sum_{i \in S}(x_i + v_i(y_i)) \mid ((x_i, y_i))_{i \in S} \in Y_S\} = \max_{z \in V(S)} \sum_{i \in S} z_i.$$

For the reverse inclusion, take $\zeta \in \{z \in \mathbb{R}^S \mid \sum_{i \in S} z_i \leq v(S)\}$ and let $((0, y_i))_{i \in S} \in Y_S$ be such that $\sum_{i \in S} v_i(y_i) = \max\{\sum_{i \in S} v_i(\tilde{y}_i) \mid ((\hat{x}_i, \tilde{y}_i))_{i \in S} \in Y_S\} = v(S)$. Next, define $x_i = \zeta_i - v_i(y_i)$ for all agents $i \in S$. Then

$$\begin{aligned} \sum_{i \in S} x_i &= \sum_{i \in S} \zeta_i - v_i(y_i) \\ &\leq v(S) - \max\left\{\sum_{i \in S} v_i(\tilde{y}_i) \mid ((0, \tilde{y}_i))_{i \in S} \in Y_S\right\} \\ &= 0, \end{aligned}$$

and, consequently, $((x_i, y_i))_{i \in S} \in Y_S$. Since $U_i(x_i, y_i) = x_i + v_i(y_i) = \zeta_i$ for all $i \in S$ it follows that $\{z \in \mathbb{R}^S \mid \sum_{i \in S} z_i \leq v(S)\} \subset V(S)$.

Summarizing, if the utility functions are linearly separable in money we can model a cooperative decision-making problem as a TU-game.

Next, let us turn to the second question, that is, which preferences can be represented by a utility function that is linearly separable in money. For characterizing these preferences we need the properties (P1), (P2) and three additional properties. So, let $\succsim_i$ be a preference relation on an outcome space $Y = \mathbb{R} \times \tilde{Y}$.

(P4) The preferences $\succsim_i$ are called *strictly monotonic* in x if for any outcomes $(x, y), (\tilde{x}, y) \in \mathbb{R} \times \tilde{Y}$ it holds that $(x, y) \succ_i (\tilde{x}, y)$ if and only if $x > \tilde{x}$.

(P5) For any outcomes $(x, y), (\tilde{x}, \tilde{y}) \in \mathbb{R} \times \tilde{Y}$ with $(x, y) \succsim_i (\tilde{x}, \tilde{y})$ there exists $t \in \mathbb{R}$ such that $(x, y) \sim_i (\tilde{x} + t, \tilde{y})$.

(P6) For any outcomes $(x, y), (\tilde{x}, \tilde{y}) \in \mathbb{R} \times \tilde{Y}$ with $(x, y) \sim_i (\tilde{x}, \tilde{y})$ and every $t \in \mathbb{R}$ it holds that $(x + t, y) \sim_i (\tilde{x} + t, \tilde{y})$.

Property (P4) means that agent i prefers more money to less. Property (P5) states that a change in the outcome from (x, y) to $(\tilde{x}, \tilde{y})$ can be compensated by an amount of money t. Finally, property (P6) implies that agent i's preferences over the outcome space $\tilde{Y}$ does not depend on the amount of money x_i he receives.

The following theorem is due to Aumann (1960) and Kaneko (1976) and characterizes preference relations with transferable utility.

Theorem 2.5 Let $\succsim_i$ be a preference relation on the outcome space $Y = \mathbb{R} \times \tilde{Y}$. If $\succsim_i$ satisfies conditions (P1), (P2), and (P4) - (P6), then there exists a utility function $v_i : \tilde{Y} \to \mathbb{R}$ such that for any $(x, y), (\hat{x}, \hat{y}) \in Y$ it holds that $(x, y) \succsim_i (\hat{x}, \hat{y})$ if and only if $x + v_i(y) \geq \hat{x} + v_i(\hat{y})$.

Remark that if U_i represents the preferences $\succsim_i$, then so does any monotonic transformation of U_i. The property of linear separability in money, however, may be lost after such monotonic transformations. For instance, suppose $U_i(x, y) = x + v_i(y)$ represents agent i's preferences $\succsim_i$ on $\mathbb{R} \times \tilde{Y}$. The utility function $\hat{U}_i$ defined by $\hat{U}_i(x, y) = e^{U_i(x,y)}$ for all $(x, y) \in \mathbb{R} \times \tilde{Y}$ also represents $\succsim_i$. The utility function $\hat{U}_i$, however, is not linearly separable in money since $\hat{U}_i(x, y) = e^{x+v_i(y)}$ for all $(x, y) \in \mathbb{R} \times \tilde{Y}$.

2.3 Cooperative Games with Transferable Utility

A *cooperative game with transferable utility*, or *TU-game*, is described by a pair (N, v), where $N \subset \mathbb{N}$ is the set of agents and $v : 2^N \to \mathbb{R}$ is the *characteristic function* assigning to each coalition $S \subset N$ a value $v(S)$, representing the benefits from cooperation. In particular, we have that $v(\emptyset) = 0$.

Example 2.6 For the linear production situation formulated in Example 2.2 we have the outcome space $Y = \mathbb{R}$ and for each $S \subset N$

$$Y_S = \{(x_i)_{i\in S} \in \mathbb{R}^S | \exists_{c\in C(S)} : \sum_{i\in S} x_i = p^\top c\}$$

with $C(S) = \{c \in \mathbb{R}^M_+ | Ac \leq \sum_{i\in S} b^i\}$ the set of feasible production plans. Since the outcome space Y represents allocations of money only, transferable utility implies that $U_i(x) = x$ for all $x \in \mathbb{R}$ and all $i \in N$. A linear production game (N, v) as introduced by Owen (1975), is then defined by

$$v(S) = \max\{\sum_{i\in S} U_i(x_i) | (x_i)_{i\in S} \in Y_S\} = \max\left\{p^\top c | Ac \leq \sum_{i\in S} b^i\right\}$$

for all $S \subset N$.

Example 2.7 Recall that for the sequencing situation described in Example 2.3 we have $N \subset \mathbb{N}$, the set of divisions waiting for service,

$$Y = \mathbb{R} \times \Pi_N,$$

the outcome space, and

$$Y_S = \{((x_i, \sigma))_{i\in S} \in \prod_{i\in S} Y | \sum_{i\in S} x_i \leq 0 \text{ and } \sigma \in \Pi_S\}$$

the outcome space for coalition S, where Π_S denotes the set of admissible rearrangements with respect to the initial serving order σ_0. Let the waiting costs for division $i \in N$ be given by $k_i(\sigma) = \sum_{j\in N:\sigma(j)\leq\sigma(i)} \alpha_i p_j$ with p_j the service time of division j and $\alpha_j > 0$, i.e. the waiting costs increase linearly with the waiting time. Then given a serving order σ the cost savings for division i equal $k_i(\sigma_0) - k_i(\sigma)$. Next, let $U_i(x, \sigma) = x + k_i(\sigma_0) - k_i(\sigma)$ for all $(x, \sigma) \in Y_S$. So, given an outcome (x, σ) division i's utility equals the amount of money x it receives plus the cost savings that correspond with the serving order σ. A sequencing game (N, v) as introduced by Curiel, Pederzoli and Tijs (1989), is then defined by

$$\begin{aligned} v(S) &= \max\left\{\sum_{i\in S} U_i(x_i, \sigma) | ((x_i, \sigma))_{i\in S} \in Y_S\right\} \\ &= \max\left\{\sum_{i\in S} (k_i(\sigma_0) - k_i(\sigma)) | \sigma \in \Pi_S\right\}, \end{aligned}$$

for all $S \subset N$.

A TU-game (N, v) is called *superadditive* if for all disjoint $S, T \subset N$ it holds that

$$v(S) + v(T) \leq v(S \cup T). \tag{2.4}$$

This means that two disjoint coalitions do not suffer from forming one large coalition. Many economic situations will lead to superadditive games, for if two disjoint coalitions S and T join forces, they can often guarantee themselves the payoff $v(S) + v(T)$ by operating separately.

A TU-game (N, v) is *convex* if for all $i \in N$ and all $S \subset T \subset N \backslash \{i\}$ it holds that

$$v(S \cup \{i\}) - v(S) \leq v(T \cup \{i\}) - v(T). \tag{2.5}$$

Convexity thus implies that an agent contributes more to the benefits of a coalition when this coalition becomes larger. Convexity first appeared in Shapley (1971), and besides the definition given by (2.5), several other, equivalent definitions exist. Two of those are the following. The first one is quite similar to (2.5) and reads as follows: for each $U \subset N$ and each $S \subset T \subset N \backslash U$ it holds that

$$v(S \cup U) - v(S) \leq v(T \cup U) - v(T). \tag{2.6}$$

So, also if more than one agent joins a coalition, the change in the benefits increases when this coalition increases. The second alternative definition is significantly different from expression (2.5); it states that for every $S, T \subset N$ we have that

$$v(S) + v(T) \leq v(S \cap T) + v(S \cup T). \tag{2.7}$$

One can show that expression (2.7) implies that $v(S \cup T) - v(S) - v(T)$ is increasing in both S and T. Since $v(S \cup T) - v(S) - v(T)$ can be interpreted as the incentive for coalitions S and T to merge, convexity thus means that larger coalitions have a bigger incentive to join forces. Hence, convexity is a much stronger condition than superadditivity, which only states that two disjoint coalitions have an incentive to merge; superadditivity does not say anything about the magnitude of this incentive. That convex games are superadditive also follows immediately from (2.7) by taking S and T disjoint. Convexity, however, is not as generally satisfied as superadditivity. The linear production games described in Example 2.6, for instance, are superadditive but not necessarily convex.

What we are interested in are 'fair' allocations of the benefits $v(N)$. So, assuming that all agents are willing to cooperate with each other, how should the benefits $v(N)$ be divided so that each agent is satisfied. The game theoretical literature proposes many allocation rules as a solution to this problem. As we will show, some of these rules are generally applicable while others are designed to suit only very specific TU-games. Each of these rules, of course, has it own pros and cons and deciding which allocation rule applies best is a problem in itself.

An *allocation* of $v(N)$ is described by a vector $x \in \mathbb{R}^N$ such that $\sum_{i\in N} x_i \leq v(N)$. Moreover, an allocation x is called *efficient* or *Pareto optimal* if there exists no allocation $\hat{x}$ that yields every agent a higher payoff, i.e., $\hat{x}_i > x_i$ for all $i \in N$. Obviously, each agent i is only willing to participate in the grand coalition if it pays him at least as much as he can obtain on his own. This means that an allocation x must be such that $x_i \geq v(\{i\})$ for all $i \in N$. These allocations are called *individually rational*. The set of all individually rational and Pareto optimal allocations is called the *imputation set* and is defined by

$$I(v) = \{x \in \mathbb{R}^N |\ \sum_{i\in N} x_i = v(N), \forall_{i\in N} : x_i \geq v(\{i\})\}. \tag{2.8}$$

Note that the imputation set is nonempty if the game is superadditive.

The argument that an agent does not participate in the grand coalition N if he does not receive at least $v(\{i\})$ is reasonable. Moreover, it does not only apply to individuals, but also to coalitions. A coalition S has incentives to part company with the grand coalition if it can improve the payoff of each member. Mathematically, this means that given an allocation x of $v(N)$, coalition S has incentives to leave coalition N if there exists an allocation y of $v(S)$ such that $y_i > x_i$ for all $i \in S$. It is a straightforward exercise to check that such an allocation y does not exist if and only if $\sum_{i\in S} x_i \geq v(S)$. Hence, no coalition has incentives to separate from the grand coalition if the allocation x is such that $\sum_{i\in S} x_i \geq v(S)$ for all $S \subset N$. Such an allocation x is called a *core-allocation*. The *core* of a TU-game (N, v) is thus defined by

$$C(v) = \{x \in I(v) |\ \forall_{S\subset N} : \sum_{i\in S} x_i \geq v(S)\}. \tag{2.9}$$

Although core-allocations provide the agents with an incentive to maintain the grand coalition, they need not always exist, as the following example shows.

Example 2.8 Consider a three-person TU-game (N, v) with $v(\{i\}) = 0$ for all $i \in N$ and $v(S) = 1$ for all $S \subset N$ with $|S| \geq 2$. Note that this game is superadditive but that its core is empty. To see that no core allocation exists, take $x \in I(v)$. So, $\sum_{i\in N} x_i = 1$ and $x_i \geq 0$ for $i = 1, 2, 3$. If x would be a core-allocation it must satisfy $x_1 + x_2 \geq 1$, $x_1 + x_3 \geq 1$, and $x_2 + x_3 \geq 1$. Adding these three inequalities yields $2(x_1 + x_2 + x_3) \geq 3$. Since $x_1 + x_2 + x_3 = 1$ this inequality cannot be satisfied. Hence, the core is empty.

A necessary and sufficient condition for nonemptiness of the core is given in Bondareva (1963) and Shapley (1967). In order to formulate this condition we use the notion of balanced maps. Therefore, define for each $S \subset N$ the vector $e^S \in \mathbb{R}^N$ by $e_i^S = 1$ if $i \in S$ and $e_i^S = 0$ if $i \notin S$. A map $\lambda : 2^N \to [0,1]$ is a *balanced map* if for all $i \in N$ it holds that $\sum_{S \subset N} \lambda(S) e^S = e^N$.

Theorem 2.9 Let (N, v) be a TU-game. Then $C(v) \neq \emptyset$ if and only if for each balanced map λ it holds that

$$\sum_{S \subset N} \lambda(S) v(S) \leq v(N). \tag{2.10}$$

TU-games that have a nonempty core are called *balanced games*. Furthermore, if also every subgame $(S, v_{|S})$ of the TU-game (N, v) has a nonempty core, the game is called *totally balanced*. Here, the characteristic function of a subgame $(S, v_{|S})$ is defined as $v_{|S}(T) = v(T)$ for all $T \subset S$. The linear production games and sequencing games presented in Example 2.6 and Example 2.7, respectively, are examples of totally balanced games.

A sufficient but not necessary condition for nonemptiness of the core is provided in Shapley (1971) and reads as follows:

Theorem 2.10 Let (N, v) be a TU-game. If (N, v) is convex, then $C(v) \neq \emptyset$.

When dividing the benefits of the grand coalition N, it is likely that the agents focus on core-allocations only. For if an allocation outside the core is proposed, at least one coalition can threaten to leave the grand coalition if it does not receive a larger share of the benefits. The core, however, need not give a decisive answer to what allocation the agents should agree upon. Although the core can consist of only one allocation, most of the time it turns out to be a fairly large set of allocations. For illustration, consider the following example.

Example 2.11 Consider a three-person TU-game (N, v) with $v(\{i\}) = 0$ for all $i \in N$, $v(\{1,2\}) = 4$, $v(\{1,3\} = 3$, $v(\{2,3\} = 2$, and $v(\{1,2,3\}) = 8$. The core $C(v)$ of this game is depicted in Figure 2.4 and contains infinitely many allocations.

In contrast with the core, which can be a set of allocations, an allocation rule assigns to each TU-game (N, v) exactly one allocation. A well known allocation rule is the Shapley value, introduced in Shapley (1953). This allocation rule is based on the marginal vectors of a TU-game (N, v). To explain the definition of a marginal vector, let σ be an ordering of the agents. So, $\sigma : N \to \{1, 2, \ldots, n\}$

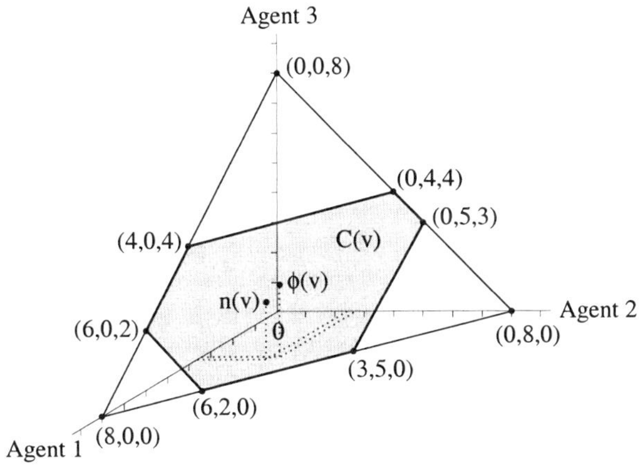

Figure 2.4

is a bijection with $\sigma(i) = j$ meaning that agent i is in position j. Next, let the agents enter a room one at a time according to the order σ. Thus, agent $\sigma^{-1}(1)$ enters first, then agent $\sigma^{-1}(2)$, and so on. When an agent enters the room, he joins the coalition that is already there. Furthermore, the payoff he receives equals the amount with which the benefits of the coalition changes. The resulting allocation is called the *marginal vector* with respect to the order σ and equals

$$m_i^\sigma(v) = v(\{j \in N | \sigma(j) \leq \sigma(i)\}) - v(\{j \in N | \sigma(j) < \sigma(i)\}), \quad (2.11)$$

for all $i \in N$. Note that $m^\sigma(v)$ is indeed an allocation since $\sum_{i \in N} m_i^\sigma(v) = v(N)$. Presuming that each of the $n!$ different orders occurs equally likely, the *Shapley value* $\phi(v)$ assigns to each agent i his expected marginal contribution, that is,

$$\begin{aligned} \phi_i(v) &= \frac{1}{n!} \sum_{\sigma \in \Pi_N} m_i^\sigma(v) \quad (2.12) \\ &= \sum_{S \subset N \setminus \{i\}} \frac{\#S!(n-1-\#S)!}{n!} (v(S \cup \{i\}) - v(S)), \end{aligned}$$

with $\Pi_N = \{\sigma : N \to \{1, 2, \ldots, n\} | \forall_{i,j \in N : i \neq j} : \sigma(i) \neq \sigma(j)\}$ the set of all orderings of N. Let us calculate the Shapley value for the game given in Example 2.11.

Example 2.12 Consider the game defined in Example 2.11. Let σ_i, $i = 1, 2, \ldots, 6$ denote the orders 123, 132, 312, 321, 231, and 213, respectively. Then $m_1^{\sigma_1}(v) =$

$v(\{1\})-v(\emptyset)=0, m_2^{\sigma_1}(v)=v(\{1,2\})-v(\{1\})=4$, and $m_3^{\sigma_1}(v)=v(\{1,2,3\})-v(\{2,3\})=4$. So, $m^{\sigma_1}(v)=(0,4,4)$. Similarly, one calculates $m^{\sigma_2}(v)=(0,5,3)$, $m^{\sigma_3}(v)=(3,5,0)$, $m^{\sigma_4}(v)=(6,2,0)$, $m^{\sigma_5}(v)=(6,0,2)$, and $m^{\sigma_6}(v)=(4,0,4)$. The Shapley value thus equals $\phi(v)=\frac{1}{6}\sum_{i=1}^{6}m^{\sigma_i}(v)=\frac{1}{6}(19,16,13)$ (see also Figure 2.4).

As can be seen in Figure 2.4, the Shapley value is a core-allocation for this particular game. In general, however, this is not the case as the following example shows.

Example 2.13 Consider the TU game (N,v), with $N=\{1,2,3\}$ and $v(S)=1$ if $S\in\{\{1,3\},\{2,3\},\{1,2,3\}\}$ and zero otherwise. Note that $c(v)=\{(0,0,1)\}$. To calculate the Shapley value, let σ_i, $i=1,2,\ldots,6$ denote the orders 123, 132, 312, 321, 231, and 213, respectively. Then $m_1^{\sigma_1}(v)=v(\{1\})-v(\emptyset)=0$, $m_2^{\sigma_1}(v)=v(\{1,2\})-v(\{1\})=0$, and $m_3^{\sigma_1}(v)=v(\{1,2,3\})-v(\{1,2\})=1$. So, $m^{\sigma_1}(v)=(0,0,1)$. Similarly, one calculates $m^{\sigma_2}(v)=(0,0,1)$, $m^{\sigma_3}(v)=(1,0,0)$, $m^{\sigma_4}(v)=(0,1,0)$, $m^{\sigma_5}(v)=(0,0,1)$, and $m^{\sigma_6}(v)=(0,0,1)$. The Shapley value then equals $\phi(v)=\frac{1}{6}\sum_{i=1}^{6}m^{\sigma_i}(v)=\frac{1}{6}(1,1,4)$. Note that $\phi(v)$ is not a core-allocation since $\phi_1(v)+\phi_2(v)=\frac{5}{6}<1=v(\{1,2\})$.

So, the Shapley value need not result in a core-allocation. Shapley (1971) provides a sufficient condition for $\phi(v)$ to be an element of the core.

Theorem 2.14 Let (N,v) be a TU-game. If (N,v) is convex, then $\phi(v)\in C(v)$.

The proof of this proposition follows from a result also due to Shapley (1971), namely, that for convex games all marginal vectors belong to the core. The converse of this statement is given in Ichiishi (1981). Hence, a TU-game (N,v) is convex if and only if all marginal vectors are elements of the core $C(v)$.

As opposed to the Shapley value, the nucleolus is an allocation rule that results in a core-allocation whenever the core is nonempty. The nucleolus for TU-games is introduced in Schmeidler (1969) and minimizes the maximal complaint of the coalitions. Given an allocation $x\in I(v)$, the complaint of a coalition S expresses how dissatisfied this coalition is with the proposed allocation x. The complaint of coalition S is defined as the difference between what this coalition can obtain on its own and what the allocation x assigns to the members of S. So, the less x assigns to the members of S, the higher the complaint of this coalition will be. Then, roughly speaking, the nucleolus is that allocation of the imputation set $I(v)$ that minimizes the maximal complaint of all coalitions.

For a formal definition of the nucleolus, let (N, v) be a TU-game such that the imputation set $I(v)$ is nonempty. Given an allocation x, the complaint, or excess, of coalition S is defined by $E(S,x) = v(S) - \sum_{i \in S} x_i$. Note that the excess $E(S,x)$ increases if $\sum_{i \in S} x_i$ decreases. Next, let $E(x) = (E(S,x))_{S \subset N}$ be the vector of excesses and let $\theta \circ E(x)$ denote the vector of excesses with its elements arranged in decreasing order. The *nucleolus* is then defined by

$$n(v) \;=\; \{x \in I(v) |\, \forall_{y \in I(v)} : \theta \circ E(x) \leq_{lex} \theta \circ E(y)\}. \tag{2.13}$$

Here, $\leq_{lex}$ denotes the lexicographical ordering on $\mathbb{R}^{2^N}$. Given two vectors $x, y \in \mathbb{R}^{2^N}$ we have $x <_{lex} y$ if there exists $k \in \mathbb{N}$ such that $x_i = y_i$ for $i = 1, 2, \ldots, k$ and $x_{k+1} < y_{k+1}$. So, when comparing two vectors, the lexicographical ordering first compares the first element of each vector, if they are equal it compares the second element of each vector, and so on. The following result is due to Schmeidler (1969).

Theorem 2.15 Let (N, v) be a TU-game. If $I(v) \neq \emptyset$, then the nucleolus $n(v)$ is a singleton. Moreover, $n(v) \subset C(v)$ whenever $C(v) \neq \emptyset$.

Example 2.16 Consider again the game presented in Example 2.11. The nucleolus for this game equals $n(v) = (3\frac{1}{2}, 2\frac{1}{2}, 2)$ with the maximal complaint being $E(\{3\}, n(v)) = E(\{1,2\}, n(v)) = -2$. Figure 2.4 shows that the nucleolus $n(v)$ is indeed a core-allocation.

Although the nucleolus has the nice property that it always results in a core-allocation, provided one exists, it is not that easy to calculate. Even for a three player example the calculations are extensive, hence the absence of any calculations in Example 2.16. There are special classes of TU-games though, for which the nucleolus is determined more easily. Examples of such TU-games are big boss games (see Muto, Nakayama, Potters and Tijs (1988)) and airport games (see Littlechild (1974)).

The two allocation rules discussed thus far are applicable to every (superadditive) TU-game. When considering a particular class of TU-games though, a more intuitive allocation rule may exist that applies to this particular setting only. Besides having a more appealing interpretation, these rules are often easier to determine than both the Shapley value and the nucleolus. This, for instance, is the case for linear production games and sequencing games. We therefore conclude this section on TU-games with two examples of such specific allocation rules, one applying to linear production games and the other applying to sequencing games.

Example 2.17 Recall that a linear production game (N, v) is defined by

$$v(S) = \max\left\{p^\top c \mid Ac \leq \sum_{i\in S} b^i, c \geq 0\right\}$$

for all $S \subset N$. One can, of course, calculate the Shapley value and nucleolus for these games. Since linear production games are totally balanced, we even know that the nucleolus results in a core-allocation. The following, relatively easy procedure, however, also yields a core-allocation for every linear production game (N, v).

Consider the following linear program to determine the value $v(N)$ of the grand coalition

$$v(N) = \max\left\{p^\top c \mid Ac \leq \sum_{i\in N} b^i, c \geq 0\right\}.$$

The dual of this linear program is

$$\min\left\{y^\top \sum_{i\in N} b^i \mid y^\top A \geq p, y \geq 0\right\}.$$

Now, if $\hat{y} \in \mathbb{R}^r_+$ denotes an optimal solution of the dual program, Owen (1975) shows that the allocation x defined by $x_i = \hat{y}^\top b^i$ is a core-allocation of the linear production game (N, v). For interpreting this allocation, note that the optimal solution $\hat{y}$ represents the shadow prices of the resources. This means that if, say, the amount of resource j increase with 1 unit, then the maximal revenue increases with approximately $\hat{y}_j$. In other words, $\hat{y}_j \cdot \sum_{i\in N} b^i_j$ denotes the contribution of resource j in the maximal revenues. The allocation $(\hat{y}^\top b^i)_{i\in \mathrm{N}}$ then gives each agent the value of his resource bundle.

Example 2.18 For the sequencing games described in Example 2.7, Curiel et al. (1989) introduced the Equal Gain Splitting rule. For understanding how this rule works, let us go into more detail on the optimal serving order. For this purpose, consider two divisions i and j such that division j is the immediate follower of division i. Recall that the waiting costs for divisions i and j are given by $k_i(\sigma) = \sum_{k\in N:\sigma(k)\leq\sigma(i)} \alpha_i p_k$ and $k_j(\sigma) = \sum_{k\in N:\sigma(k)\leq\sigma(j)} \alpha_j p_k$, respectively. Now, if division i and j change their positions in the serving order, then division i's waiting time increases with p_j. Hence, the waiting costs for division i increase with $\alpha_i p_j$. Similarly, the waiting time for division j decreases with p_i, so that its waiting costs decrease with $\alpha_j p_i$. The benefits from this change in position thus equal $\alpha_j p_i - \alpha_i p_j$. This implies that division i and j trade places if and only if the benefits are positive, that is, $\alpha_j p_i - \alpha_i p_j > 0$. It can be shown that an optimal order is reached by consecutively switching the places of two divisions, one succeeding the other, whenever such a change is beneficial. Moreover, all beneficial switches occur in this procedure. The *Equal Gain Splitting rule*, or EGS-rule, then divides the benefits of each switch equally among the two divisions that are involved in this switch. Hence, division i receives

$$EGS_i(v) = \sum_{\substack{j\in N:\ \sigma_0(j)>\sigma_0(i)\\ \alpha_j p_i - \alpha_i p_j > 0}} \tfrac{1}{2}(\alpha_j p_i - \alpha_i p_j) + \sum_{\substack{k\in N:\ \sigma_0(k)<\sigma_0(i)\\ \alpha_i p_k - \alpha_k p_i > 0}} \tfrac{1}{2}(\alpha_i p_k - \alpha_k p_i).$$

Curiel et al. (1989) shows that the EGS-rule always results in a core-allocation of the corresponding sequencing game. Other allocation rules arise when the benefits $\alpha_j p_i - \alpha_i p_j$ are not divided equally among the divisions i and j, but according to some distribution code $\gamma = \{\gamma_{ij}\}_{i,j\in N:i<j}$. This means that if divisions i and j change their positions, division i receives the fraction $\gamma_{ij} \in [0,1]$ of the benefits $\alpha_j p_i - \alpha_i p_j$ and division j receives the fraction $1-\gamma_{ij} \in [0,1]$. These allocation rules are considered in Hamers, Suijs, Borm and Tijs (1996). For an extensive study on (generalized) sequencing situations we refer to Hamers (1996).

2.4 Cooperative Games with Non-Transferable Utility

A *cooperative game with non-transferable utility*, or *NTU-game*, is described by a pair (N, V), where $N = \{1, 2, \ldots, n\}$ denotes the set of agents and V is a map assigning to each coalition S a nonempty and closed subset $V(S)$ of $\mathbb{R}^S$ such that

$$V(\{i\}) = (-\infty, v(\{i\})],\ v(\{i\}) \in \mathbb{R}, \tag{2.14}$$

for all $i \in N$,

$$\text{if } x \in V(S) \text{ and } \hat{x} \leq x \text{ then also } \hat{x} \in V(S), \tag{2.15}$$

for all $S \subset N$, and

$$V(S) \cap \{x \in \mathbb{R}^S |\ \forall_{i\in S} : x_i \geq v(\{i\})\} \tag{2.16}$$

is bounded for all $S \subset N$. The set $V(S)$ represents the utility vectors the members of coalition S can obtain when cooperating (see also (2.1)). For mathematical reasons it is sometimes useful to extend the utility vectors in $V(S)$ with zeros for the agents that do not belong to coalition S. So, let us also define

$$\hat{V}(S) = \{x \in \mathbb{R}^N |\ (x_i)_{i\in S} \in V(S), \forall_{i\notin S} : x_i = 0\} \tag{2.17}$$

for all $S \subset N$.

Example 2.19 Every TU-game (N, v) can be modelled as an NTU-game (N, V) by defining

$$V(S) = \{x \in \mathbb{R}^S \mid \sum_{i \in S} x_i \leq v(S)\}$$

for all $S \subset N$.

Example 2.20 Consider the model of an exchange economy presented in Example 2.1. Assuming that the agents' preferences can be represented by a continuous utility function U_i, this situation gives rise to the following NTU-game (N, V) with

$$V(S) = \{z \in \mathbb{R}^S \mid \exists_{(c^i)_{i \in S} \in Y_S} \forall_{i \in S} : z_i \leq U_i(c^i)\},$$

for all $i \in S$. Note that since Y_S is compact and U_i is continuous for all $i \in N$, condition (2.16) is satisfied. These games are known in the literature as market games and were analyzed in e.g. Scarf (1967) and Shapley and Shubik (1969).

An NTU-game (N, V) is called *superadditive* if for all disjoint $S, T \subset N$ we have that

$$V(S) \times V(T) \subset V(S \cup T). \tag{2.18}$$

Similar as for TU-games, superadditivity implies that two disjoint coalitions do not suffer from merging into one large coalition. Usually, superadditivity is satisfied. For two disjoint coalitions S and T can at least generate the benefits $V(S) + V(T)$ by pretending a merger and, in the meantime, operating separately. An equivalent statement for superadditivity is that $\tilde{V}(S) \times \tilde{V}(T) \subset \tilde{V}(S \cup T)$ for all disjoint $S, T \subset N$.

An NTU-game (N, V) is called *ordinal convex* if for all $S, T \subset N$ and all $x \in \mathbb{R}^N$ the following statement is true: if $(x_i)_{i \in S} \in V(S)$ and $(x_i)_{i \in T} \in V(T)$ then either $(x_i)_{i \in S \cap T} \in V(S \cap T)$ or $(x_i)_{i \in S \cup T} \in V(S \cup T)$. Ordinal convexity is introduced in Vilkov (1977) and extends the definition of convexity as given in (2.7) to NTU-games. Another such an extension is cardinal convexity, introduced in Sharkey (1981). For defining cardinal convexity, let (N, V) be an NTU-game such that $V(\{i\}) = (-\infty, 0]$ for all $i \in N$. Then (N, V) is *cardinal convex* if for all $S, T \subset N$ it holds that

$$\tilde{V}(S) + \tilde{V}(T) \subset \tilde{V}(S \cup T) + \tilde{V}(S \cap T). \tag{2.19}$$

Note that $V(\{i\}) = (-\infty, 0]$ is not really a restriction because of the monotonic transformations we may take from the utility functions. Ordinal and cardinal convexity are not equivalent though, Sharkey (1981) provides two examples of NTU-games, one being ordinal but not cardinal convex and the other being cardinal but not ordinal convex.

As was the case for TU-games, we assume that the grand coalition N is formed and focus on 'fair' allocations, that is, utility vectors $x \in V(N)$ at which all agents are satisfied. In this context, an *allocation* for coalition S is a utility vector $x \in V(S)$. An allocation $x \in V(S)$ is called *efficient* or *Pareto optimal* if there exists no other allocation $\hat{x} \in V(S)$ that yields all agents in S a higher utility, i.e., $\hat{x}_i > x_i$ for all $i \in S$. Furthermore, an agent will only participate in the grand coalition if this yields him as least as much as he can obtain on his own. This means that only those allocations $x \in V(N)$ are eligible such that for each agent $i \in N$ it holds that $x_i \geq v(\{i\})$. These allocations are called *individually rational*. Furthermore, the set of all efficient and individually rational allocations is called the *imputation set* and is given by

$$I(V) = \{x \in V(N) |\ \nexists_{\hat{x} \in V(N)} \forall_{i \in N} : \hat{x}_i > x_i, \forall_{i \in N} : x_i \geq v(\{i\})\}. \quad (2.20)$$

Analogous to TU-games one can define the core of an NTU-game. A core-allocation is such that no coalition S can improve the payoff of each member by separating from the grand coalition N and operating on its own. A coalition S can improve upon the allocation $x \in V(N)$ if there exists an allocation $\hat{x} \in V(S)$ such that $\hat{x}_i > x_i$ for all $i \in S$. The *core* of an NTU-game is thus given by

$$C(V) = \{x \in V(N) |\ \forall_{S \subset N} \nexists_{\hat{x} \in V(S)} \forall_{i \in S} : \hat{x}_i > x_i\}. \quad (2.21)$$

Example 2.21 Consider the following two person market game with two consumption goods. Let $\omega^1 = (2, 3)$ and $\omega^2 = (4, 1)$ be the endowments of agents 1 and 2, respectively. Next, let $c^i = (c^i_1, c^i_2)$ denote the consumption bundle for agent i. The preferences for agent 1 are described by the utility function $U_1(c^1_1, c^1_2) = 2\sqrt{c^1_1} + \sqrt{c^1_2}$ and the preferences for agent 2 are described by $U_2(c^2_1, c^2_2) = \sqrt{c^2_1} + 2\sqrt{c^2_2}$. The corresponding NTU-game equals

$$\begin{aligned} V(\{1\}) &= \{x \in \mathbb{R} |\ x \leq U_1(2, 3) = 2\sqrt{2} + \sqrt{3}\} \\ V(\{2\}) &= \{x \in \mathbb{R} |\ x \leq U_2(4, 1) = 4\} \\ V(\{1, 2\}) &= \{x \in \mathbb{R}^2 |\ \exists_{c^1, c^2 \in \mathbb{R}^2 : (c^1, c^2) \leq (6,4)} : x_i \leq U_i(c^i), i = 1, 2\} \end{aligned}$$

and is depicted in Figure 2.5. As one can see, the core $C(V)$ of this game is nonempty and equals $I(V)$.

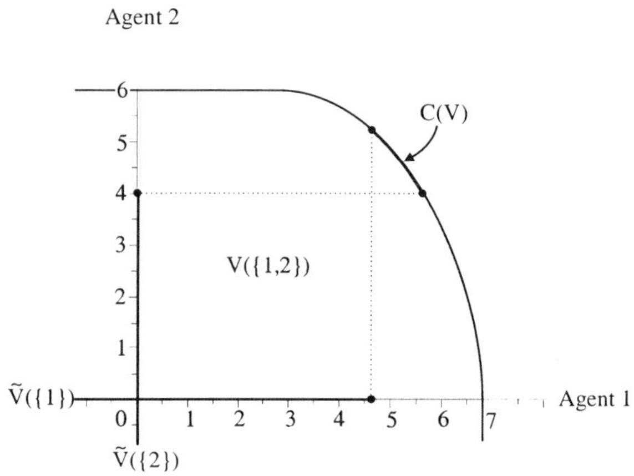

Figure 2.5

From Example 2.19 we know that every TU-game can be modelled as an NTU-game. Since the core of a TU-game consists of the same utility vectors as the core of the corresponding NTU-game, it follows from Example 2.8 that also for NTU-games the core can be empty. Furthermore, a necessary and sufficient condition like the balancedness condition for TU-games does not exist. Although a notion of balancedness exists, it only provides a sufficient condition for nonemptiness of the core. The following result is a consequence of the main theorem in Scarf (1967).

Theorem 2.22 Let (N, V) be an NTU-game. If for all balanced maps λ (see page 24) it holds that $\sum_{S \subset N} \lambda(S)\tilde{V}(S) \subset \tilde{V}(N)$, then $C(V) \neq \emptyset$.

As was the case for TU-games convexity is also a sufficient condition for a nonempty core. The following two results can be found in Vilkov (1977) and Sharkey (1981), respectively.

Theorem 2.23 Let (N, V) be an NTU-game. If (N, V) is ordinally convex, then $C(V) \neq \emptyset$.

Theorem 2.24 Let (N, V) be an NTU-game. If (N, V) is cardinally convex, then $C(V) \neq \emptyset$.

An NTU-game with a nonempty core is called *balanced*. Furthermore, if for every subgame $(S, V_{|S})$ the core is nonempty, then the game is called *totally balanced*. The subgame $(S, V_{|S})$ is defined by $V_{|S}(T) = V(T)$ for all $T \subset S$. Market games, for example, are totally balanced NTU-games.

As shown above, properties like superadditivity, convexity, and balancedness extend reasonably well to the class of NTU-games. The same can be said for the Shapley value, as opposed to the nucleolus which has yet to be defined for NTU-games. For the Shapley value several extensions have been introduced. For example, the Harsanyi-value (Harsanyi (1963)), the Shapley NTU-value (Shapley (1969)), the egalitarian solution (Kalai and Samet (1985)), the consistent Shapley value (Maschler and Owen (1992)), and the marginal based compromise value (Otten, Peleg, Borm and Tijs (1998)). In the remainder of this section we confine ourselves to a brief discussion of the Shapley NTU-value and the marginal based compromise value only.

Since the Shapley value was introduced in the previous section as the average of all marginal vectors, a straightforward extension to the class of NTU-games would thus be averaging the marginal vectors. Unfortunately, this procedure does not work. Although marginal vectors can be defined, the average of all these allocations can fail Pareto optimality. So, when extending the Shapley value to NTU-games, one has to employ a different approach.

Let us start with explaining Shapley's approach that led to the Shapley NTU-value for an NTU-game (N, V). For this purpose, assume that for each allocation $x \in V(S)$ the total utility for coalition S can be represented by a weighted sum of the individual utility levels. So, given a weight vector $w \in \mathbb{R}_+^N$ with $\sum_{i \in N} w_i = 1$, the total weighted utility that coalition S assigns to the allocation x equals $\sum_{i \in S} w_i x_i$. If we also assume that these weighted utilities are transferable, each weight vector w gives rise to a TU-game (N, v_w) with

$$v_w(S) = \sup\{\sum_{i \in S} w_i x_i | \, x \in V(S)\}, \tag{2.22}$$

for all $S \subset N$. For these games we can determine the Shapley value $\phi(v_w)$. Since the Shapley value $\phi(v_w)$ represents an allocation of the weighted sum of utilities, the resulting allocation need not be attainable for coalition N. So, what we are interested in are the attainable allocations $x \in V(N)$ for which the vector of weighted utilities $(w_i x_i)_{i \in N}$ coincides with $\phi(v_w)$. The *Shapley NTU- value* is thus defined by

$$\Phi(V) = \{x \in V(N) | \, \exists_{w \in \Delta^N} \forall_{i \in N} : \phi_i(v_w) = w_i x_i\}, \tag{2.23}$$

where $\Delta^N = \{w \in \mathbb{R}_+^N | \sum_{i \in N} w_i = 1\}$.

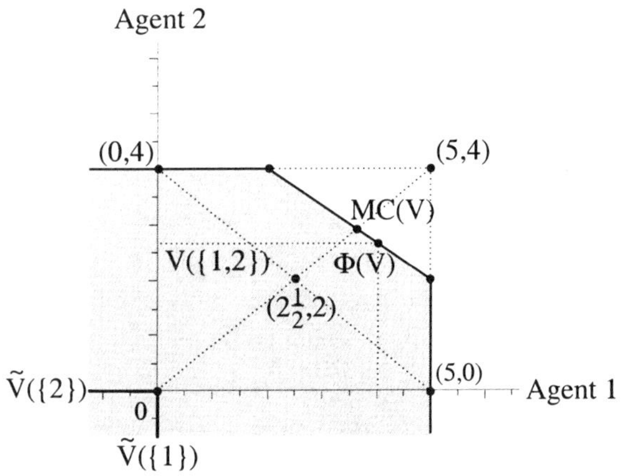

Figure 2.6

Example 2.25 Consider the following two-person NTU-game (N, V) with $V(\{i\}) = (-\infty, 0]$ for $i = 1, 2$ and $V(\{1,2\})$ as depicted in Figure 2.6. Let $w = (w_1, w_2)$ be a weight vector. Then for all $w \in \Delta^N$ it holds that $v_w(\{1\}) = v_w(\{2\}) = 0$ and

$$v_w(\{1,2\}) = \begin{cases} 4 - 2w_1, & \text{if } w_1 \leq \frac{2}{5} \\ 2 + 3w_1, & \text{if } w_1 > \frac{2}{5}. \end{cases}$$

This yields

$$\phi(v_w) = \begin{cases} (2 - w_1, 2 - w_1), & \text{if } w_1 \leq \frac{2}{5} \\ (1 + \frac{3}{2}w_1, 1 + \frac{3}{2}w_1), & \text{if } w_1 > \frac{2}{5}. \end{cases}$$

Next, let $w = (\frac{2}{5}, \frac{3}{5})$ and consider the allocation $(4, \frac{8}{3}) \in V(\{1,2\})$. The corresponding weighted utilities equal $\frac{2}{5} \cdot 4 = \frac{8}{5}$ for agent 1 and $\frac{3}{5} \cdot \frac{8}{3} = \frac{8}{5}$ for agent 2. Since this coincides with $\phi(v_w)$ it holds that $(\frac{8}{5}, \frac{8}{5}) \in \Phi(V)$.

Example 2.26 Consider the two-person NTU-game (N, V) depicted in Figure 2.7. So, $V(\{i\}) = (-\infty, 2]$ for $i = 1, 2$ and

$$V(\{1,2\}) = \{(x_1, x_2) \in \mathbb{R}^2 | \, (x_1, x_2) \leq (2,4)\} \cup \{(x_1, x_2) \in \mathbb{R}^2 | \, (x_1, x_2) \leq (4,2)\}.$$

For this game it holds that $\Phi(V) = \emptyset$. To see this, let $w = (w_1, w_2)$ be a weightvector. Then $v_w(\{1\}) = 2w_1$, $v_w(\{2\}) = 2w_2 = 2(1 - w_1)$, and

$$v_w(\{1,2\}) = \begin{cases} 4 - 2w_1, & \text{if } w_1 \leq \frac{1}{2} \\ 2 + 2w_1, & \text{if } w_1 > \frac{1}{2} \end{cases}$$

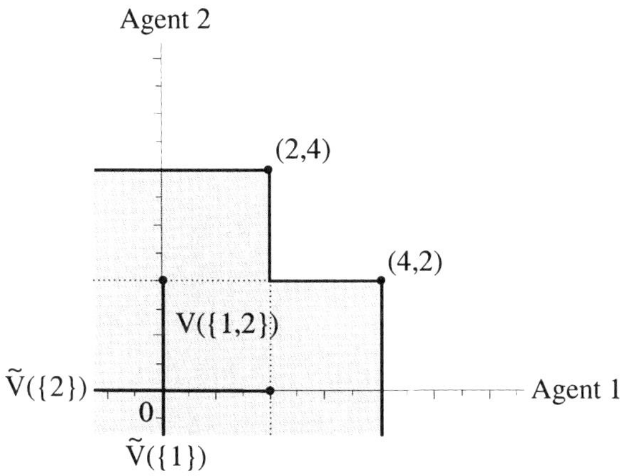

Figure 2.7

Hence, the Shapley value $\phi(v_w)$ equals

$$\phi(v_w) = \begin{cases} (1+w_1, 3-3w_1), & \text{if } w_1 \le \frac{1}{2} \\ (3w_1, 2-w_1), & \text{if } w_1 > \frac{1}{2} \end{cases}$$

Now, if $w_1 = 0$ then $\phi(v_w) = (1,3)$. Since $w_1 = 0$ there exists no $x \in V(\{1,2\})$ such that $1 = w_1 x_1$ and $3 = w_2 x_2$. If $w_1 \in (0, \frac{1}{2})$, the only candidate to belong to $\Phi(V)$ is the allocation $(2,4)$. But if $(2,4) \in \Phi(V)$ it must hold that $\phi_1(v_w) = 2w_1$ and $\phi_2(v_w) = 4w_2 = 4(1-w_1)$. Since both equations imply that $w_1 = 1$, we have a contradiction. Similar contradictions follow if $w_1 \in [\frac{1}{2}, 1]$, hence $\Phi(V) = \emptyset$.

As Example 2.26 shows, the Shapley NTU-value need not always exist. Shapley (1969) provides sufficient conditions for nonemptiness. It does, however, not guarantee unicity of the Shapley NTU-value.

Theorem 2.27 Let (N, V) be a superadditive NTU-game. If $V(N)$ is a convex subset of $\mathbb{R}^N$ then $\Phi(V) \neq \emptyset$.

As follows from the definition of the Shapley NTU-value, the marginal vectors of the NTU-game itself are not taken into account. The marginal based compromise value, as its name implies, does make use of these marginal vectors. For defining a marginal vector of an NTU-game (N, V), let $\sigma \in \Pi_N$ be an ordering of the agents. Similar to TU-games, we assign to each agent i his marginal contribution to the benefits of the coalition that is formed by the agents preceding agent i in the order σ. The *marginal vector* $m^\sigma(V)$ is thus defined recursively as follows. Let i_1 be the first agent in the order σ, that is $\sigma(i_1) = 1$. He receives

$$m^\sigma_{i_1}(V) = \sup\{x_{i_1} | \, x \in V(\{i_1\})\}.$$

The second agent i_2 then receives

$$m^\sigma_{i_2}(V) = \sup\{x_{i_2} | \, x \in V(\{i_1, i_2\}) : x_{i_1} = m^\sigma_{i_1}(V)\}.$$

Continuing this procedure yields agent i_k a payoff equal to

$$m^\sigma_{i_k}(V) = \sup\{x_{i_k} | \, x \in V(\{i_1, i_2, \ldots, i_k\}) : \forall_{j=1,2,\ldots,k-1} : x_{i_j} = m^\sigma_{i_j}(V)\}. \tag{2.24}$$

Note that a marginal vector $m^\sigma(V)$ is well defined if the game is superadditive. Furthermore, note that by definition $m^\sigma(V) \in V(N)$ and that $m^\sigma(V)$ satisfies Pareto optimality. Moreover, if this procedure is applied to TU-games the original marginal vectors result.

The marginal based compromise value is defined for NTU-games (N, V) with $V(\{i\}) = (-\infty, 0]$ for all $i \in N$. Let $\overline{m}(V) = \sum_{\sigma \in \Pi_N} m^\sigma(V)$. Then the *marginal based compromise value*, or *MC-value*, is defined by

$$MC(V) = \alpha_V \overline{m}(V), \tag{2.25}$$

where a_V is such that $\alpha_V = \sup\{\alpha \in \mathbb{R}_+ | \alpha \overline{m}(V) \in V(N)\}$. So, the MC-value is the largest multiple of the vector $\overline{m}(V)$ that still belongs to $V(N)$. Hence, it is Pareto optimal. Furthermore, the MC-value coincides with the Shapley value on the class of TU-games and with the consistent Shapley value on the class of so-called hyperplane games (see Maschler and Owen (1989)).

Example 2.28 Consider the game illustrated in Figure 2.6. Since it is a two-person game, there are only two different marginal vectors. Let σ_1 denote the order 12 and σ_2 the order 21. Then $m^{\sigma_1}_1(V) = \sup\{x | x \in V(\{1\})\} = 0$ and $m^{\sigma_1}_2(V) = \sup\{x_2 | (x_1, x_2) \in V(\{1, 2\}) : x_1 = 0\} = 4$. Similarly, $m^{\sigma_2}_2(V) = \sup\{x | x \in V(\{2\})\} = 0$ and $m^{\sigma_2}_1(V) = \sup\{x_1 | (x_1, x_2) \in V(\{1, 2\}) : x_2 = 0\} = 5$. This implies that $\overline{m}(V) = (0, 4) + (5, 0) = (5, 4)$ and that $MC(V) = (3\frac{7}{11}, 2\frac{10}{11})$. Finally, note that the average of the marginal allocations equals $(2\frac{1}{2}, 2)$ and that it violates Pareto optimality since it is in the interior of $V(N)$.

2.5 Chance-Constrained Games

Chance-constrained games as introduced by Charnes and Granot (1973) extend the theory of cooperative games in characteristic function form to situations where the benefits from cooperation are random variables. So, when several agents decide to cooperate, they do not exactly know the benefits this cooperation generates. What they do know is the probability distribution function of these benefits. Let $\boldsymbol{V}(S)$

denote the random variable describing the benefits of coalition S. Furthermore, denote its probability distribution function by $F_{\boldsymbol{V}(S)}$. Thus,

$$F_{\boldsymbol{V}(S)}(t) = \mathbb{P}\{\boldsymbol{V}(S) \leq t\}, \tag{2.26}$$

for all $t \in \mathbb{R}$. Then a *chance-constrained game* is defined by the pair $(N, \boldsymbol{V})$, with $\boldsymbol{V}$ the characteristic function assigning to each coalition S the nonnegative random benefits $\boldsymbol{V}(S)$. Note that chance-constrained games are based on the formulation (N, v) of TU-games with the deterministic benefits $v(S)$ replaced by stochastic benefits $\boldsymbol{V}(S)$. This, however, does not imply that the preferences of the agents are also linearly separable in money. In fact, the individual preferences are of no account in this model.

For dividing the benefits of the grand coalition, the authors propose two-stage allocations. In the first stage, when the realization of the benefits is still unknown, each agent is promised a certain payoff. These so-called prior payoffs are such that there is a fair chance that they are realized. Once the benefits are known, the total payoff allocated in the prior payoff can differ from what is actually available. In that case, we come to the second stage and modify the prior payoff in accordance with the realized benefits.

Let us start with discussing the prior allocations. A prior payoff is denoted by a vector $x \in \mathbb{R}^N$, with the interpretation that agent $i \in N$ receives the amount x_i. To comply with the condition that there is a reasonable probability that the promised payoffs can be kept, the prior payoff x must be such that

$$\underline{\alpha}(N) \leq \mathbb{P}(\{\boldsymbol{V}(N) \leq \sum_{i \in N} x_i\}) = F_{\boldsymbol{V}(N)}(\sum_{i \in N} x_i) \leq \overline{\alpha}(N), \tag{2.27}$$

with $0 < \underline{\alpha}(N) \leq \overline{\alpha}(N) < 1$. This condition assures that the total amount $\sum_{i \in N} x_i$ that is allocated is not too low or too high. Note that expression (2.27) can also be written as

$$\xi_{\underline{\alpha}(N)}(\boldsymbol{V}(N)) \leq \sum_{i \in N} x_i \leq \xi_{\overline{\alpha}(N)}(\boldsymbol{V}(N)).$$

To come to a prior core for chance-constrained games, one needs to specify when a coalition S is satisfied with the amount $\sum_{i \in S} x_i$ it receives, so that it does not threaten to leave the grand coalition N. Charnes and Granot (1973) assume that a coalition S is satisfied with what it gets, if the probability that they can obtain more on their own is small enough. This means that for each coalition $S \neq N$ there exists a number $\alpha(S) \in (0, 1)$ such that coalition S is willing to participate in the coalition N whenever $\mathbb{P}(\{\boldsymbol{V}(S) \leq \sum_{i \in S} x_i\}) \geq \alpha(S)$. The number $\alpha(S)$ is a measure of assurance for coalition S. Note that the measure of assurance may vary over the coalitions. Furthermore, they reflect the coalition's attitude towards risk, the willingness to bargain with other coalitions, and so on. The *prior core* of a chance-constrained game $(N, \boldsymbol{V})$ is then defined by

$$C(\boldsymbol{V}) = \left\{ x \in \mathbb{R}^N \;\middle|\; \begin{array}{l} \forall_{S \subset N: S \neq N} : F_{\boldsymbol{V}(S)}(\sum_{i \in S} x_i) \geq \alpha(S), \\ \underline{\alpha}(N) \leq F_{\boldsymbol{V}(N)}(\sum_{i \in N} x_i) \leq \overline{\alpha}(N) \end{array} \right\}. \tag{2.28}$$

Example 2.29 Consider the following three-person chance-constrained game $(N, \boldsymbol{V})$ with $\boldsymbol{V}(\{i\}) = 0$, $i = 1, 2, 3$, $\boldsymbol{V}(S) \sim U(0, 2)$ if $\#S = 2$, and $\boldsymbol{V}(N) \sim U(1, 2)$. Furthermore, let $\alpha(S) = \frac{1}{2}$ for all $S \neq N$, $\underline{\alpha}(N) = \frac{2}{5}$, and $\overline{\alpha}(N) = \frac{3}{5}$. Next, let $x \in \mathbb{R}^3$ be a prior allocation. Since

$$F_{\boldsymbol{V}(N)}(t)) = \begin{cases} 0, & \text{if } t \leq 1, \\ t - 1, & \text{if } 1 < t < 2, \\ 1, & \text{if } 2 \leq t, \end{cases}$$

condition (2.27) implies that $1\frac{2}{5} \leq x(N) \leq 1\frac{3}{5}$. For the one-person coalitions we have that $F_{\boldsymbol{V}(\{i\})}(x_i) \geq \frac{1}{2}$ if $x_i \geq 0$. Furthermore, since for all two-person coalitions S it holds that

$$F_{\boldsymbol{V}(S)}(t)) = \begin{cases} 0, & \text{if } t \leq 0, \\ \frac{1}{2}t, & \text{if } 0 < t < 2, \\ 1, & \text{if } 2 \leq t, \end{cases}$$

it follows that $\sum_{i \in S} x_i \geq 1$ if $F_{\boldsymbol{V}(S)}(\sum_{i \in S} x_i) \geq \frac{1}{2}$. Hence, the prior core of this game is given by

$$C(\boldsymbol{V}) = \{x \in \mathbb{R}^3_+ | \forall_{S \subset N: \#S = 2} : \sum_{i \in S} x_i \geq 1, 1\tfrac{2}{5} \leq \sum_{i \in N} x_i \leq 1\tfrac{3}{5}\}.$$

A necessary and sufficient condition for nonemptiness of the core is given by the following theorem, which can be found in Charnes and Granot (1973).

Theorem 2.30 Let $(N, \boldsymbol{V})$ be a chance-constrained game. Then $C(\boldsymbol{V}) = \emptyset$ if and only if $\mu > \overline{\alpha}(N)$ with

$$\begin{aligned} \mu = \; & \min \; F_{\boldsymbol{V}(N))}(\textstyle\sum_{i \in N} x_i) \\ & \text{s.t.:} \; F_{\boldsymbol{V}(S)}(\textstyle\sum_{i \in S} x_i) \geq \alpha(S) \\ & \qquad F_{\boldsymbol{V}(N)}(\textstyle\sum_{i \in N} x_i) \geq \overline{\alpha}(N). \end{aligned}$$

A prior Shapley value is defined in the same way as for TU-games. The agents enter a room one at a time and, subsequently, receive their contribution to the benefits. The prior Shapley value then assigns to each agent the expected payoff of this procedure. Thus, given a chance-constrained game $(N, \boldsymbol{V})$, the *prior Shapley value* $\phi(\boldsymbol{V})$ is defined by

$$\phi_i(\boldsymbol{V}) = \sum_{S \subset N \setminus \{i\}} \tfrac{\#S!(n-1-\#S)!}{n!} \left(E(\boldsymbol{V}(S \cup \{i\})) - E(\boldsymbol{V}(S))\right), \qquad (2.29)$$

for all $i \in N$.

Note that the prior-Shapley value extends the Shapley value for TU-games. This follows immediately from the fact that $E(v(S)) = v(S)$ in the deterministic case.

Example 2.31 Consider the game defined in Example 2.29. Since $E(\boldsymbol{V}(\{i\})) = 0$ for $i = 1, 2, 3$, $E(\boldsymbol{V}(S)) = \frac{1}{2}$ if $\#S = 2$, and $E(\boldsymbol{V}(N)) = 1\frac{1}{2}$, it follows that

$$\begin{aligned}\phi_1(\boldsymbol{V}) &= \tfrac{1}{6}\,(2(E(\boldsymbol{V}(\{1\})) - E(\boldsymbol{V}(\emptyset))) + E(\boldsymbol{V}(\{1,2\})) \\ &\quad -E(\boldsymbol{V}(\{2\})) + E(\boldsymbol{V}(\{1,3\})) - E(\boldsymbol{V}(\{3\})) \\ &\quad +2(E\boldsymbol{V}(\{1,2,3\})) - E(\boldsymbol{V}(\{2,3\})))) \\ &= \tfrac{1}{6}(2 \cdot (0-0) + 2 \cdot (\tfrac{1}{2} - 0) + 2 \cdot (1\tfrac{1}{2} - \tfrac{1}{2})) \\ &= \tfrac{1}{2}.\end{aligned}$$

Similarly, it follows that $\phi_2(\boldsymbol{V}) = \frac{1}{2}$ and $\phi_3(\boldsymbol{V}) = \frac{1}{2}$. Note that $\phi(\boldsymbol{V}) \in C(\boldsymbol{V})$.

For defining a prior nucleolus, one needs to specify the excess of a coalition at a given allocation x. Applying the definition used for TU-games results in the excess $E(S,x) = \boldsymbol{V}(S) - \sum_{i \in S} x_i$. Since $\boldsymbol{V}(S)$ is a stochastic variable, the excess is also a stochastic variable. But this raises a problem, because for determining the nucleolus we need to arrange the excesses in a decreasing order. For stochastic variables, however, there is no straightforward criterium that states when one stochastic variable is larger than another one. So, for chance-constrained games another definition of the excess is needed. Charnes and Granot (1976) express the excess of coalition S at an allocation x by $1 - F_{\boldsymbol{V}(S)}(\sum_{i \in S} x_i)$, the probability that $\boldsymbol{V}(S)$ exceeds $\sum_{i \in S} x_i$. This implies that the excess decreases with $\sum_{i \in S} x_i$. With this definition the prior nucleolus is defined similarly to the nucleolus for TU-games. This means that for each feasible allocation $x \in Y$, where

$$Y = \{x \in \mathbb{R}^N \,|\, \forall_{i \in N} : x_i \geq 0, \underline{\alpha}(N) \leq F_{\boldsymbol{V}(N)}(\sum_{i \in N} x_i) \leq \overline{\alpha}(N)\},$$

we have $E(S,x) = 1 - F_{\boldsymbol{V}(S)}(\sum_{i \in S} x_i)$ for all $S \subset N$. Furthermore, $E(x) = (E(S,x))_{S \subset N}$ is the vector of excesses and $\theta \circ E(x)$ denotes the vector of excesses with its elements arranged in decreasing order. The *prior nucleolus* of a chance-constrained game is then defined by

$$n(\boldsymbol{V}) = \{x \in Y \mid \forall_{y \in Y} : \theta \circ E(x) \leq_{lex} \theta \circ E(y)\} \tag{2.30}$$

The prior-nucleolus is not an extension of the nucleolus for TU-games. For if the benefits are deterministic, the excess $E(S,x) = 1 - F_{\boldsymbol{V}(S)}(\sum_{i \in S} x_i)$ equals either zero or one, depending on whether $\sum_{i \in S} x_i \geq v(S)$ or $\sum_{i \in S} x_i < v(S)$.

Example 2.32 Consider again the game presented in Example 2.29. Then for each feasible allocation x we have that $E(\{i\}, x) = 0$ for $i = 1,2,3$, $E(S,x) = 1 - \frac{1}{2}\sum_{i \in S} x_i$ if $\#S = 2$ and $E(N,x) = 2 - \sum_{i \in N} x_i$. The prior nucleolus of this game equals $n(\boldsymbol{V}) = \frac{1}{15}(8,8,8)$. Note that $n(\boldsymbol{V}) \in C(\boldsymbol{V})$ and that $n_i(\boldsymbol{V}) > \phi_i(\boldsymbol{V})$ for $i = 1,2,3$.

Charnes and Granot (1976) show that the prior nucleolus is a well defined allocation. The prior nucleolus, however, need not be in the prior core, as we show in the following example.

Example 2.33 Consider the following three-person chance-constrained game $(N, \boldsymbol{V})$ with $\boldsymbol{V}(\{i\}) = 0$, $i = 1,2,3$, $\boldsymbol{V}(S) \sim U(0,2)$ if $|S| \geq 2$. Furthermore, let $\alpha(\{i\}) = \frac{1}{2}$ for $i = 1,2,3$, $\alpha(\{1,2\}) = \frac{1}{8}$, $\alpha(\{1,3\}) = \frac{2}{8}$, $\alpha(\{2,3\}) = \frac{3}{8}$, and $\underline{\alpha}(N) = \overline{\alpha}(N) = \frac{3}{8}$. Then $x \in \mathbb{R}^3_+$ is an element of the prior core if

$$\begin{array}{rrrcl} x_1+ & x_2+ & x_3 & = & \frac{3}{4} \\ x_1+ & x_2 & & \geq & \frac{1}{4} \\ x_1+ & & x_3 & \geq & \frac{1}{2} \\ & x_2+ & x_3 & \geq & \frac{3}{4}. \end{array}$$

The only solution of this system of inequalities is $x = (0, \frac{1}{4}, \frac{1}{2})$. Hence, $C(\boldsymbol{V}) = \{(0, \frac{1}{4}, \frac{1}{2})\}$. For determining the nucleolus, note that for any feasible allocation x it holds that $E(\{i\}, x) = 0$ for $i = 1,2,3$ and that $E(S,x) = 1 - \frac{1}{2}x(S)$ for $\#S = 2$. One can check that the prior nucleolus then equals $n(\boldsymbol{V}) = \frac{1}{4}(1,1,1)$, which is not contained in the prior core.

Thus far, we only discussed first stage allocations. As mentioned before, once the realization of the benefits is known, the first stage allocation might need some modifications. For these modifications, Charnes and Granot (1977) introduce the two-stage nucleolus. Given the prior allocation x, for instance, the prior Shapley value, and the realization $\hat{v}(N)$ of $\boldsymbol{V}(N)$, the two-stage nucleolus allocates $\hat{v}(N)$ in such a way that the maximal complaint of the coalitions is minimized. An allocation of $\hat{v}(N)$ is a vector $y \in \hat{Y} = \{\hat{y} \in \mathbb{R}^n_+ | \sum_{i \in N} \hat{y}_i = \hat{v}(N)\}$. Given an allocation $y \in \hat{Y}$ the complaint of coalition S with respect to this modification is defined by

$$\begin{aligned} E(S, y; x) &= \mathbb{P}(\{\boldsymbol{V}(S) > \textstyle\sum_{i \in S} y_i\}) - \mathbb{P}(\{\boldsymbol{V}(S) > \textstyle\sum_{i \in S} x_i\}) \\ &= F_{\boldsymbol{V}(S)}(\textstyle\sum_{i \in S} x_i) - F_{\boldsymbol{V}(S)}(\textstyle\sum_{i \in S} y_i). \end{aligned}$$

So, the complaint increases if $\sum_{i \in S} y_i$ decreases and the other way around. Now, if $E(y; x) = (E(S, y; x))_{S \subset N}$ denotes the vector of excesses and $\theta \circ E(y; x)$ the vector of excesses with the elements ordered in a decreasing way, the *two-stage nucleolus* with respect to the prior allocation x is given by

$$\hat{n}(\boldsymbol{V}; x) = \{y \in \hat{Y} | \forall_{\hat{y} \in \hat{Y}} : \theta \circ E(y; x) \leq_{lex} \theta \circ E(\hat{y}; x)\}. \tag{2.31}$$

Example 2.34 Consider the game defined in Example 2.29. Recall that the prior Shapley value equals $\phi(\boldsymbol{V}) = \frac{1}{2}(1, 1, 1)$. Now, suppose that the realization of $\boldsymbol{V}(N)$ is 1, then the prior-allocation needs to be modified. Let $\hat{Y} = \{\hat{y} \in \mathbb{R}^3_+ | \sum_{i \in N} \hat{y}_i = 1\}$. Furthermore, for $y \in \hat{Y}$ we have that $E(\{i\}, y; \phi(\boldsymbol{V})) = 0$ for $i = 1, 2, 3$, $E(S, y; \phi(\boldsymbol{V})) = \frac{1}{2}(\sum_{i \in S} \phi_i(\boldsymbol{V}) - \sum_{i \in S} y_i)$ if $\#S = 2$, and $E(N, y; \phi(\boldsymbol{V})) = 1\frac{1}{2} - \sum_{i \in N} y_i$. This implies that $\hat{n}(\boldsymbol{V}; \phi(\boldsymbol{V})) = \frac{1}{3}(1, 1, 1)$.

3 Stochastic Cooperative Games

For TU-games, the payoff of a coalition is assumed to be known with certainty. In many cases though, the payoffs to coalitions can be uncertain. This would not raise a problem if the agents can await the realizations of the payoffs before deciding which coalitions to form and which allocations to settle on. But if the formation of coalitions and allocations has to take place before the payoffs are realized, the theory of TU-games does no longer apply.

Section 2.5 of the previous chapter discussed chance-constrained games that are introduced in Charnes and Granot (1973). For these games the value of a coalition S is allowed to be a random variable. They suggested to allocate the stochastic payoff of the grand coalition in two stages. In the first stage, so-called prior payoffs are promised to the agents. These prior payoffs are such that there is a good chance that these promises can be realized. In the second stage the realization of the stochastic payoff is awaited and, subsequently, a possibly nonfeasible prior payoff vector has to be adjusted to this realization in some way.

In this chapter we present the basic model of stochastic cooperative games as introduced by Suijs et al. (1999) . In contrast to chance-constrained games, stochastic cooperative games explicitly include the agents' preferences. This enables us to take into account each individual's behavior towards risk. Furthermore, it allocates random payoffs to each agent instead of the two-stage deterministic allocations in chance-constrained games.

3.1 The Model

For a better understanding of the definition of a stochastic cooperative game, let us start with describing some examples of cooperative decision-making problems in a stochastic environment. So, let us return to the linear production and sequencing situations and formulate them in a stochastic setting.

Example 3.1 Recall from Example 2.2 that for the deterministic case the outcome space Y in a linear production situation equals $\mathbb{R}$ and that for each coalition S it holds that

$$Y_S = \{(x_i)_{i\in S} \in \mathbb{R}^S | \exists_{c\in C(S)} : \sum_{i\in S} x_i = p^\top c\},$$

with $C(S) = \{c \in \mathbb{R}_+^M | Ac \leq \sum_{i\in S} b^i\}$ the set of feasible production plans. As already mentioned in the introduction, we can formulate this problem in a stochastic setting. Presuming that production takes a considerable amount of time, prices can change between the moment that a decision is taken on the production plan and the moment that this production plan is realized. So, the individuals do not exactly know the prices when they have to decide on their production plan. Let the stochastic variable $\boldsymbol{P}_j \in L^1(\mathbb{R})$ describe the price of commodity $j \in M$. Then given a feasible production plan $c \in C(S)$ for coalition S, the revenues equal $\boldsymbol{P}^\top c$ with $\boldsymbol{P} = (\boldsymbol{P}_j)_{j\in M}$. Since the revenues are stochastic, each individual receives a stochastic payoff so that we can choose the outcome space $\mathcal{Y}$ to be $L^1(\mathbb{R})$. Furthermore, an allocation of the revenues $\boldsymbol{P}^\top c$ to the members of coalition S is a vector $\boldsymbol{Y} \in \prod_{i\in S} \boldsymbol{Y} = L^1(\mathbb{R})^S$ such that $\sum_{i\in S} \boldsymbol{Y}_i \leq \boldsymbol{P}^\top c$. The outcome space $\mathcal{Y}_S$ thus equals

$$\mathcal{Y}_S = \{(\boldsymbol{Y}_i)_{i\in S} \in L^1(\mathbb{R})^S | \exists_{c\in C(S)} : \sum_{i\in S} \boldsymbol{Y}_i \leq \boldsymbol{P}^\top c\}. \tag{3.1}$$

If $\succsim_i$ describes the preferences of agent $i \in N$ over the set of random payoffs $L^1(\mathbb{R})$, then the triple $(N, \{\mathcal{Y}_S\}_{S\subset N}, \{\succsim_i\}_{i\in N})$ describes a linear production situation with stochastic prices.

Example 3.2 Consider the sequencing situation described in Example 2.3. In the more realistic case, the service times of the divisions are only known with certainty once the service has ended. As long as they are waiting for service, the divisions only know their serving times by approximation. So, let the stochastic variable $\boldsymbol{P}_i \in L^1(\mathbb{R})$ describe the service time of division i. Furthermore, let $k_i : \mathbb{R} \to \mathbb{R}_+$ be the cost function of division i. Thus, division i incurs costs $k_i(t)$ if the waiting time equals t. Then, given a serving order $\sigma \in \Pi_N$, the total cost savings equal

$$\sum_{i\in N} k_i \left(\sum_{j\in N:\sigma_0(j)\leq\sigma_0(i)} \boldsymbol{P}_j \right) - \sum_{i\in N} k_i \left(\sum_{j\in N:\sigma(j)\leq\sigma(i)} \boldsymbol{P}_j \right).$$

Since these cost savings are stochastic the outcome space can be chosen $\mathcal{Y} = L^1(\mathbb{R})$. Similarly as in Example 3.1 one defines

$$\mathcal{Y}_S = \{(\boldsymbol{Y}_i)_{i\in S} \in L^1(\mathbb{R})^S | \exists_{\sigma\in\Pi_S} : \sum_{i\in S} \boldsymbol{Y}_i \leq \boldsymbol{K}_S(\sigma)\}, \tag{3.2}$$

with $\boldsymbol{K}_S(\sigma) = \sum_{i\in S} k_i \left(\sum_{j\in N:\sigma_0(j)\leq\sigma_0(i)} \boldsymbol{P}_j\right) - \sum_{i\in S} k_i \left(\sum_{j\in N:\sigma(j)\leq\sigma(i)} \boldsymbol{P}_j\right)$ the random cost savings for coalition S. Note that the serving order is no longer explicitly contained in the outcome space, as is the case in Example 2.3. A sequencing situation with stochastic processing times can thus be described by $(N, \{\mathcal{Y}_S\}_{S \subset N}, \{\succsim_i\}_{i \in N})$, where $\succsim_i$ are agent i's preferences over the set $L^1(\mathbb{R})$ of random payoffs.

The third and final example concerns financial markets. For a general equilibrium model on financial markets the reader is referred to Magill and Shafer (1991). The example we provide will show some substantial differences with the model considered by Magill and Shafer (1991). First, our example focuses on cooperation between the agents, and second, the assets we consider may be indivisible goods.

Example 3.3 Let N be a set of agents, each having an initial endowment ω^i of money. Furthermore, we have a set M of financial assets, where each financial asset $j \in M$ has a price π_j and stochastic revenues $\boldsymbol{R}_j \in L^1(\mathbb{R})$. Each agent can invest his money in a portfolio of assets and obtain stochastic revenues. We allow the set M to contain identical assets, so that we do not need to specify the amount that each agent buys of a specific asset. Instead of buying portfolios individually, agents can also cooperate with each other, combine their endowments of money, and invest in a more diversified portfolio of assets. Although cooperation enables them to make their investment less risky, they also have to agree on a division of the joint revenues. To model this as a cooperative decision-making model, note that the agents only need to divide the stochastic revenues of their portfolio, so that we can again choose the outcome space $\mathcal{Y}$ to be $L^1(\mathbb{R})$. Next, let $A \subset M$ be a portfolio of assets. Then a portfolio A is affordable for coalition S if they can pay for all the assets in the portfolio, that is, $\sum_{j \in A} \pi_j \leq \sum_{i \in S} \omega^i$. Let $A(S) = \{A \subset F \,|\, \sum_{j \in A} \pi_j \leq \sum_{i \in S} \omega^i\}$ denote the set of affordable portfolios for coalition S. The outcome space $\mathcal{Y}_S$ then equals

$$\mathcal{Y}_S = \{(\boldsymbol{Y}_i)_{i \in S} \in L^1(\mathbb{R})^S \,|\, \exists_{A \in A(S)} : \sum_{i \in S} \boldsymbol{Y}_i \leq \sum_{j \in A} \boldsymbol{R}_j\}. \tag{3.3}$$

In the above mentioned examples coalitions have to choose between several actions, with each action possibly yielding different stochastic revenues. In other words, each coalition can choose their benefits from a collection $\mathcal{V}(S) \subset L^1(\mathbb{R})$ of random variables. This observation leads to the following definition of a stochastic cooperative game.

A *stochastic cooperative game* is described by a tuple $(N, \mathcal{V}, \{\succsim_i\}_{i \in N})$, where $N = \{1, 2, \ldots, n\}$ is the set of agents, $\mathcal{V}$ a map assigning to each coalition S a collection of stochastic payoffs $\mathcal{V}(S)$, and $\succsim_i$ the preference relation of agent i over the set of random payoffs $L^1(\mathbb{R})$. So, in particular, it is assumed that the random payoffs are expressed in some infinitely divisible commodity like money. Benefits consisting of several different or indivisible commodities are excluded.

Besides the preferences, note that the collection of random payoffs $\mathcal{V}(S)$ is another difference with a chance-constrained game, which only assigns one random payoff $\boldsymbol{V}(S)$ to each coalition. This would not be a restriction though, if it is possible to pick a 'best' stochastic payoff $\boldsymbol{V}(S) \in \mathcal{V}(S)$. For the linear production situations with stochastic prices, however, it is usually not possible to determine unambiguously the production plan that yields the best revenues. Hence, such situations cannot be modelled as chance-constrained games. The same holds for the other two examples we discussed.

Example 3.4 Consider the linear production situation with stochastic prices presented in Example 3.1. This situation is described as a stochastic cooperative game $(N, \mathcal{V}, \{\succsim_i\}_{i \in N})$ with

$$\mathcal{V}(S) = \{\boldsymbol{X} \in L^1(\mathbb{R}) |\, \exists_{c \in C(S)} : \boldsymbol{X} = \boldsymbol{P}^\top c\},$$

for each $S \subset N$.

Example 3.5 The sequencing situation with stochastic serving times as described in Example 3.2 is written as a stochastic cooperative game $(N, \mathcal{V}, \{\succsim_i\}_{i \in N})$, where

$$\mathcal{V}(S) = \{\boldsymbol{X} \in L^1(\mathbb{R}) |\, \exists_{\sigma \in \Pi_S} : \boldsymbol{X} = \boldsymbol{K}_S(\sigma)\},$$

for all $S \subset N$.

Example 3.6 The financial markets presented in Example 3.3 are described by the following stochastic cooperative game $(N, \mathcal{V}, \{\succsim_i\}_{i \in N})$ with

$$\mathcal{V}(S) = \{\boldsymbol{X} \in L^1(\mathbb{R}) |\, \exists_{A \in A(S)} : \boldsymbol{X} = \sum_{f \in A} \boldsymbol{R}_f\},$$

for all $S \subset N$.

As is the case for TU- and NTU-games, the main issue for stochastic cooperative games is to find an appropriate allocation of the stochastic payoff of the grand coalition. In Examples 3.1 - 3.3 an *allocation* of a random payoff $\boldsymbol{X}$ for a coalition S is a vector $(\boldsymbol{Y}_i)_{i \in S} \in L^1(\mathbb{R})^S$ such that $\sum_{i \in S} \boldsymbol{Y}_i \leq \boldsymbol{X}$. The interpretation is that agent $i \in S$ receives the random payoff $\boldsymbol{Y}_i$. This definition induces a very large class of allocations, which, on the one hand, is nice, but, on the other hand, will give computational difficulties. Therefore we restrict our attention to a specific class of allocations.

Let $S \subset N$ and let $\boldsymbol{X} \in \mathcal{V}(S)$ be a stochastic payoff for coalition S. An *allocation* of $\boldsymbol{X}$ can be represented by a pair $(d, r) \in \mathbb{R}^S \times \mathbb{R}^S$ with $\sum_{i \in S} d_i \leq 0$,

$\sum_{i \in S} r_i = 1$, and $r_i \geq 0$ for all $i \in S$. Given a pair (d, r), agent $i \in S$ then receives the random payoff $d_i + r_i\boldsymbol{X}$. So, an allocation consists of two parts. The first part represents deterministic transfer payments between the agents in S. Note that the $\leq$-sign allows the agents to discard some of the money. The second part then allocates a fraction of the random payoff to each agent in S. Note that we can indeed allocate fractions of the stochastic revenues $\boldsymbol{X}$ because of our assumption that $\boldsymbol{X}$ represents money or any other comparable commodity. The *class of stochastic cooperative games* with agent set N adopting this restricted definition of an allocation is denoted by $SG(N)$ and its elements are denoted by Γ. Furthermore, let $\mathcal{Z}_\Gamma(S)$ denote the set of allocations coalition S can obtain. Hence,

$$\mathcal{Z}_\Gamma(S) = \{(d_i + r_i\boldsymbol{X})_{i \in S} | \, \boldsymbol{X} \in \mathcal{V}(S), (d, r) \in H^S \times \Delta^S\}, \tag{3.4}$$

where $H^S = \{d \in \mathbb{R}^S | \sum_{i \in S} d_i \leq 0\}$ and $\Delta^S = \{r \in \mathbb{R}^S_+ | \sum_{i \in S} r_i = 1\}$.

3.2 Preferences on Stochastic Payoffs

One of the main differences between stochastic cooperative games and chance-constrained games is that the former explicitly takes into account the preferences of the individuals. Reason for this is that the behavior towards risk may vary between different individuals. In general, three different kinds of behavior are distinguished, i.e., risk averse, risk neutral, and risk loving behavior. To formalize these three types, consider the space $L^1(\mathbb{R})$ of stochastic variables and let $\succsim_i$ describe the preferences of individual i over $L^1(\mathbb{R})$. Then individual i is said to behave *risk averse* if for all $\boldsymbol{Y} \in L^1(\mathbb{R})$ it holds that $E(\boldsymbol{Y}) \succsim_i \boldsymbol{Y}$. So, he rather receives the expected payoff with certainty than the random payoff itself. Similarly, individual i is *risk loving* if the reverse holds, that is, for all $\boldsymbol{Y} \in L^1(\mathbb{R})$ it holds that $\boldsymbol{Y} \succsim_i E(\boldsymbol{Y})$. Thus he prefers receiving the random payoff to receiving its expected value. Finally, individual i is called *risk neutral* if $\boldsymbol{Y} \sim_i E(\boldsymbol{Y})$ for all $\boldsymbol{Y} \in L^1(\mathbb{R})$. This means that he is indifferent between receiving the random payoff and receiving the certain payoff $E(\boldsymbol{Y})$. So a risk neutral person is only interested in the expected payoff he receives. He does not care about the difference between the possible realizations. Note, however, that these conditions only specify whether an agent is risk averse, risk neutral, or risk loving. It does not say anything about the degree of risk aversion.

Next, let us consider some examples of preferences on stochastic variables and see how they express different kinds of risk behavior.

A natural way of ordering stochastic payoffs is by means of stochastic dominance. Let $\boldsymbol{X}, \boldsymbol{Y} \in L^1(\mathbb{R})$ be stochastic variables and denote by $F_{\boldsymbol{X}}$ and $F_{\boldsymbol{Y}}$ the probability distribution functions of $\boldsymbol{X}$ and $\boldsymbol{Y}$, respectively. Then $\boldsymbol{X}$ *stochastically dominates* $\boldsymbol{Y}$, in notation $\boldsymbol{X} \succsim_F \boldsymbol{Y}$, if and only if for all $t \in \mathbb{R}$ it holds that $F_{\boldsymbol{X}}(t) \leq F_{\boldsymbol{Y}}(t)$. Moreover, we have $\boldsymbol{X} \succ_F \boldsymbol{Y}$ if and only if for all $t \in \mathbb{R}$ it holds that $F_{\boldsymbol{X}}(t) \leq F_{\boldsymbol{Y}}(t)$ with strict inequality for at least one $t \in \mathbb{R}$. Hence, $\boldsymbol{X}$ stochastically dominates $\boldsymbol{Y}$ if for any $t \in \mathbb{R}$ the random variable $\boldsymbol{X}$ yields at most the value t with a lower probability than $\boldsymbol{Y}$ does. Note that this preference relation is incomplete. Many stochastic variables will be incomparable with respect to $\succsim_F$. Unless $\boldsymbol{Y}$ is degenerate, the random payoff $\boldsymbol{Y}$ and $E(\boldsymbol{Y})$, for instance, are incomparable. Consequently, $\succsim_F$ does not imply risk averse, risk neutral, or risk loving behavior. Intuitively though, one may expect that every rationally behaving agent, whether he is risk averse, risk neutral or risk loving, will prefer a stochastic payoff $\boldsymbol{X}$ to $\boldsymbol{Y}$ if $\boldsymbol{X} \succsim_F \boldsymbol{Y}$. So for any other preference relation $\succsim_i$ one may at least expect that $\boldsymbol{X} \succsim_i \boldsymbol{Y}$ whenever $\boldsymbol{X} \succsim_F \boldsymbol{Y}$.

In Section 2.2 we saw that under certain conditions, the preferences of an agent can be represented by a utility function. Since in that case the outcome space Y only needs to be connected, Theorem 2.4 also holds if $Y = L^1(\mathbb{R})$. So, if an agent's preferences $\succsim_i$ are complete, transitive, and continuous on $L^1(\mathbb{R})$, then there exists a continuous utility function $W_i : L^1(\mathbb{R}) \to \mathbb{R}$ such that for any $\boldsymbol{X}, \boldsymbol{Y} \in L^1(\mathbb{R})$ the following statement is true: $\boldsymbol{X} \succsim_i \boldsymbol{Y}$ if and only if $W_i(\boldsymbol{X}) \geq W_i(\boldsymbol{Y})$. Note that if the preferences in a stochastic cooperative game $\Gamma = (N, \mathcal{V}, \{\succsim_i\}_{i \in N})$ are represented by a utility function $W_i : L^1(\mathbb{R}) \to \mathbb{R}$, then we can also define an NTU-game (N, V) by

$$V(S) = \{x \in \mathbb{R}^S | \exists_{(d_i + r_i \boldsymbol{X})_{i \in S} \in \mathcal{Z}_\Gamma(S)} \forall_{i \in S} : x_i \leq W_i(d_i + r_i \boldsymbol{X})\},$$

for all $S \subset N$.

A special subclass of utility functions are the ones that correspond to expected utility. This means that there exists another utility function $U_i : \mathbb{R} \to \mathbb{R}$ such that $W_i(\boldsymbol{X}) = E(U_i(\boldsymbol{X}))$ for all $\boldsymbol{X} \in L^1(\mathbb{R})$. The utility function U_i is also called a *von Neumann-Morgenstern utility function*. It implies that individual i prefers one random payoff to another one if and only if the expected utility of the first exceeds the expected utility of the latter, that is, for any $\boldsymbol{X}, \boldsymbol{Y} \in L^1(\mathbb{R})$ it holds that $\boldsymbol{X} \succsim_i \boldsymbol{Y}$ if and only if $E(U_i(\boldsymbol{X})) \geq E(U_i(\boldsymbol{Y}))$. Agents having such preferences are called expected utility maximizers.

Note that if U_i is a von Neumann-Morgenstern utility function representing the preferences $\succsim_i$, then so does the utility function $\hat{U}_i(x) = aU_i(x) + b$ ($x \in \mathbb{R}$) with $a > 0$ and $b \in \mathbb{R}$. Note that only positive linear transformations are allowed instead of arbitrary monotonic transformations. Furthermore, note that $E(U_i(\boldsymbol{X})) \geq E(U_i(\boldsymbol{Y}))$ whenever $\boldsymbol{X}$ stochastically dominates $\boldsymbol{Y}$.

Von Neumann-Morgenstern utility functions are commonly used to model the preferences of an individual. In particular, they can express different kinds of behavior towards risk. So does risk averse behavior correspond to a concave utility function, risk neutral behavior to a linear utility function, and risk loving behavior to a convex utility function. Furthermore, if the utility function is twice differentiable, one can determine other measures of risk aversion like absolute risk aversion, relative risk aversion, and partial risk aversion, which all express a degree of risk aversion. We will not explain these terms in detail though, the interested reader is referred to Eeckhoudt and Gollier (1995).

Besides maximizing expected utility, one can think of other ways to describe a preference relation over stochastic payoffs. One such a way considers a particular quantile of the stochastic payoff. This means that for a given value of $\alpha \in (0,1)$ it holds that $\boldsymbol{X} \succsim_{\alpha} \boldsymbol{Y}$ if and only if $\xi_{\alpha}(\boldsymbol{X}) \geq \xi_{\alpha}(\boldsymbol{Y})$, were $\xi_{\alpha}(\boldsymbol{X})$ denotes the α-quantile of $\boldsymbol{X}$.[1] Furthermore, note that $\boldsymbol{X} \succsim_{\alpha} \boldsymbol{Y}$ if $\boldsymbol{X} \succsim_{F} \boldsymbol{Y}$ and that $\boldsymbol{X} \succsim_{F} \boldsymbol{Y}$ if for all $\alpha \in (0,1)$ it holds that $\boldsymbol{X} \succsim_{\alpha} \boldsymbol{Y}$. These type of preferences occur, for instance, in insurance problems when the insurance premium is based on the percentile-principle. As Example 3.7 will show, these preferences do not allow for risk averse, risk neutral, or risk loving behavior in the sense as defined on page 47. Nevertheless, a low value of α can be associated with some kind of 'risk averse' behavior. For a low value of α implies that attention is focused on the worse outcomes, which is more likely behavior for people who do not like to take risks than for people who do like it. Using a similar argument, a high value of α shows some kind of 'risk loving' behavior.

Example 3.7 Consider $\succsim_{\alpha}$-preferences with $\alpha \in (0,1)$ and define

$$\boldsymbol{X} = \begin{cases} 0, & \text{with probability } p \\ 1, & \text{with probability } 1-p, \end{cases}$$

and

$$\boldsymbol{Y} = \begin{cases} 0, & \text{with probability } q \\ 1, & \text{with probability } 1-q, \end{cases}$$

where $0 < q < \alpha < p < 1$. Then $\xi_{\alpha}(\boldsymbol{X}) = 0$ and $\xi_{\alpha}(\boldsymbol{Y}) = 1$ while $\xi_{\alpha}(E(\boldsymbol{X})) = E(\boldsymbol{X}) = 1-p$ and $\xi_{\alpha}(E(\boldsymbol{Y})) = E(\boldsymbol{Y}) = 1-q$. Hence, $E(\boldsymbol{X}) \succ_{\alpha} \boldsymbol{X}$ and $\boldsymbol{Y} \succ_{\alpha} E(\boldsymbol{Y})$, so that $\succsim_{\alpha}$ implies neither risk averse nor risk neutral nor risk loving behavior.

The third type of preferences we consider appear in portfolio decision theory. An agent's preferences over different portfolios often depends on the expected

[1] See page 129 for a formal definition.

returns of this portfolio and the variance of the returns, which is interpreted as a measure for the risk of a portfolio. Formally, this means that agent i has a utility function $U_i : \mathbb{R}^2 \to \mathbb{R}$ such that he weakly prefers the portfolio with returns $\boldsymbol{X} \in L^1(\mathbb{R})$ to a portfolio $\boldsymbol{Y}$ if and only if $U_i(E(\boldsymbol{X}), V(\boldsymbol{X})) \geq U_i(E(\boldsymbol{Y}), V(\boldsymbol{Y}))$. A simple example of such a utility function is $U_i(E(\boldsymbol{X}), V(\boldsymbol{X})) = E(\boldsymbol{X}) + b\sqrt{V(\boldsymbol{X})}$, where $b \in \mathbb{R}$. Then $b < 0$ implies risk averse behavior, $b = 0$ implies risk neutral behavior, and $b > 0$ implies risk loving behavior. This utility function, however, violates the condition we posed at the beginning of this section, namely, that $\boldsymbol{X}$ is preferred to $\boldsymbol{Y}$ if $\boldsymbol{X}$ stochastically dominates $\boldsymbol{Y}$.

Example 3.8 Given $\boldsymbol{X}, \boldsymbol{Y} \in L^1(\mathbb{R})$ let $\boldsymbol{X} \succsim \boldsymbol{Y}$ if $E(\boldsymbol{X}) - 2\sqrt{V(\boldsymbol{X})} \geq E(\boldsymbol{Y}) - 2\sqrt{V(\boldsymbol{Y})}$. Take $\boldsymbol{X} \sim U(0,2)$ and $\boldsymbol{Y} \sim U(0,1)$ so that $\boldsymbol{X}$ is stochastically dominant to $\boldsymbol{Y}$. Since $E(\boldsymbol{X}) = 1$, $V(\boldsymbol{X}) = \frac{1}{3}$, $E(\boldsymbol{Y}) = \frac{1}{2}$, and $V(\boldsymbol{Y}) = \frac{1}{12}$ it follows that $E(\boldsymbol{X}) - 2\sqrt{V(\boldsymbol{X})} = 1 - \frac{2}{3}\sqrt{3} < \frac{1}{2} - \frac{1}{3}\sqrt{3} = E(\boldsymbol{Y}) - 2\sqrt{V(\boldsymbol{Y})}$. Hence, $\boldsymbol{X} \prec \boldsymbol{Y}$.

In the forthcoming analysis of stochastic cooperative games, our attention is focused on two different subclasses of games that are characterized by the conditions we impose on the preferences of the individuals. The first class is mainly determined by a continuity property. As we will show, the standard continuity condition is too restrictive for our model. Therefore, we introduce a weaker form of continuity. For the second class, we consider preferences for which a random payoff can be represented by its so-called certainty equivalent.

3.2.1 Weakly Continuous Preferences

Consider the following three properties for the preference relation $\succsim_i$ of agent i:

(C1) $\succsim_i$ is complete and transitive on $L^1(\mathbb{R})$;

(C2) for all $\boldsymbol{X} \in L^1(\mathbb{R})$ and all $d > 0$ we have that $\boldsymbol{X} + d \succ_i \boldsymbol{X}$;

(C3) for any $\boldsymbol{X}, \boldsymbol{Y} \in L^1(\mathbb{R})$ there exist $\bar{d}, \underline{d} \in \mathbb{R}$ such that $\boldsymbol{X} + \underline{d} \prec_i \boldsymbol{Y} \prec_i \boldsymbol{X} + \bar{d}$;

(C4) $\succsim_i$ is *continuous*, i.e., the sets $\{F_{\boldsymbol{X}} \in \mathcal{F} | \boldsymbol{X} \succ_i \boldsymbol{Y}\}$ and $\{F_{\boldsymbol{X}} \in \mathcal{F} | \boldsymbol{X} \prec_i \boldsymbol{Y}\}$ are open in the metric space $(\mathcal{F}, \rho)$ (see (A.6)) for all $\boldsymbol{Y} \in L^1(\mathbb{R})$.[2]

Condition (C2) states that the preferences are strictly increasing in the deterministic amount of money one receives. Condition (C3) then states that by adding the appropriate deterministic amount of money a stochastic payoff $\boldsymbol{X}$ can be made

[2] If the preferences are complete, an equivalent statement is that $\{F_{\boldsymbol{X}} \in \mathcal{F} | \boldsymbol{X} \succsim_i \boldsymbol{Y}\}$ and $\{F_{\boldsymbol{X}} \in \mathcal{F} | \boldsymbol{X} \precsim_i \boldsymbol{Y}\}$ are closed sets in $(\mathcal{F}, \rho)$ for all $\boldsymbol{Y} \in L^1(\mathbb{R})$.

strictly worse than the stochastic payoff $\boldsymbol{Y}$ and the other way around. Finally, the continuity condition (C4) can be interpreted as follows. Given that a stochastic payoff $\boldsymbol{Y}$ is strictly preferred to the stochastic payoff $\boldsymbol{X}$, then $\boldsymbol{Y}$ is also strictly preferred to all stochastic payoffs that differ only slightly from $\boldsymbol{X}$. Note that the conditions (C1) - (C4) imply that for any $\boldsymbol{X}, \boldsymbol{Y} \in L^1(\mathbb{R})$ there exists $d \in \mathbb{R}$ such that $d + \boldsymbol{X} \sim_i \boldsymbol{Y}$. Condition (C4), however, turns out to be rather strong. As the following examples show, neither $\succsim_\alpha$-preferences nor preferences based on von Neumann-Morgenstern utility functions satisfy continuity.

Example 3.9 Consider $\succsim_\alpha$-preferences with $\alpha = \frac{1}{2}$. Take $\boldsymbol{Y} = \frac{3}{4}$ and $\boldsymbol{X}^k$ such that

$$F_{\boldsymbol{X}^k}(t) = \begin{cases} 0, & \text{if } t < 0 \\ \frac{1}{2} - \frac{1}{k} + \frac{2}{k}t, & \text{if } 0 \leq t < 1 \\ 1, & \text{if } 1 \leq t. \end{cases}$$

for all $t \in \mathbb{R}$ and $k \geq 2$. The random variable $\boldsymbol{X}^k$ attains the values 0 and 1 with probability $\frac{1}{2} - \frac{1}{k}$ each, while the remaining $\frac{2}{k}$ is equally distributed over the open interval $(0, 1)$. Then $\xi_\alpha(\boldsymbol{X}^k) = \frac{1}{2} < \frac{3}{4} = \xi_\alpha(\boldsymbol{Y})$ so that $\boldsymbol{X}^k \in \{F_{\boldsymbol{Z}} \in \mathcal{F} | \boldsymbol{Z} \precsim_i \boldsymbol{Y}\}$ for all $k \geq 2$. Since the sequence $\boldsymbol{X}^k$ weakly converges to $\boldsymbol{X}$ with $\mathbb{P}(\{\boldsymbol{X} = 0\}) = \mathbb{P}(\{\boldsymbol{X} = 1\}) = \frac{1}{2}$ it follows that $\xi_\alpha(\boldsymbol{X}) = 1$. Hence, $\boldsymbol{X} \notin \{F_{\boldsymbol{Z}} \in \mathcal{F} | \boldsymbol{Z} \precsim_i \boldsymbol{Y}\}$ so that $\succsim_\alpha$ is not continuous.

Example 3.10 Let $U_i : \mathbb{R} \to \mathbb{R}$ be a utility function defined by $U_i(t) = -e^{-t}$ for all $t \in \mathbb{R}$. Take $\boldsymbol{Y} = -\log 2$ and

$$\boldsymbol{X}^k = \begin{cases} 0, & \text{with probability } 1 - \frac{1}{k} \\ -\log(2k), & \text{with probability } \frac{1}{k} \end{cases}$$

for $k \geq 2$. Then $E(U_i(\boldsymbol{X}^k)) = (1 - \frac{1}{k})(-1) + \frac{1}{k}(-2k) = \frac{1}{k} - 3 < -2 = E(U_i(\boldsymbol{Y}))$ so that $\boldsymbol{X}^k \in \{F_{\boldsymbol{Z}} \in \mathcal{F} | \boldsymbol{Z} \precsim_i \boldsymbol{Y}\}$ for all $k \geq 2$. Next, let $\boldsymbol{X} \in L^1(\mathbb{R})$ be such that $\mathbb{P}(\boldsymbol{X} = 0) = 1$. Since $E(U_i(\boldsymbol{X})) = -1$ we have that $\boldsymbol{X} \notin \{F_{\boldsymbol{Z}} \in \mathcal{F} | \boldsymbol{Z} \precsim_i \boldsymbol{Y}\}$. From the fact that the sequence $\boldsymbol{X}^k$ weakly converges to $\boldsymbol{X} = 0$ it then follows that $\{F_{\boldsymbol{Z}} \in \mathcal{F} | \boldsymbol{Z} \precsim_i \boldsymbol{Y}\}$ is not closed. Since the preferences are complete, this implies that they are not continuous.

Note that the utility function in Example 3.10 is unbounded. For bounded utility functions, expected utility does lead to a continuous preference relation. This follows directly from Helly's Theorem A.9, that is stated in Appendix A.

As the two examples show, continuity is too strong a condition. So, if we want something like continuity to be satisfied, we need some further assumptions. Actually, only one additional assumption and a weaker form of continuity is what we need. Let $\Gamma \in SG(N)$ be a stochastic cooperative game such that $\mathcal{V}(S)$ is finite for every $S \subset N$. So, each coalition only has a finite number of random benefits to choose from. Note that this is the case for the sequencing games and financial markets presented in Example 3.5 and Example 3.6, but not for the linear production games described in Example 3.4.

For our weaker form of continuity, we make use of the special structure of the allocations. This means that we need continuity on specific subsets of random variables only. Therefore, let $\Gamma = (N, \mathcal{V}, \{\succsim_i\}_{i \in N}) \in SG(N)$ and suppose that coalition $S \subset N$ has formed. Then the set of payoffs agent $i \in S$ possibly obtains equals

$$\{d_i + r_i \boldsymbol{X} \mid \boldsymbol{X} \in \mathcal{V}(S), d_i \in \mathbb{R}, r_i \in [0,1]\}.$$

Consequently, the set of payoffs agent i possibly obtains in the game Γ equals

$$\{d_i + r_i \boldsymbol{X} \mid \exists_{S \subset N : i \in S, \# S \geq 2} : \boldsymbol{X} \in \mathcal{V}(S), d_i \in \mathbb{R}, r_i \in [0,1]\} \cup \mathcal{V}(\{i\}). \quad (3.5)$$

So, what we actually need is that the preferences $\succsim_i$ are continuous on the set defined in (3.5). Formulating continuity in this way though, implies that continuity depends on the game Γ. Since that is not desirable, we formulate it in the following, more generalized way. For this purpose, let $\mathcal{X} \subset L^1(\mathbb{R})$ be a finite subset of random variables. Define

$$\mathcal{L}(\mathcal{X}) = \{d + r\boldsymbol{X} \mid \boldsymbol{X} \in \mathcal{X}, d \in \mathbb{R}, r \in [0,1]\}, \quad (3.6)$$

and

$$\mathcal{F}(\mathcal{X}) = \{F_{\boldsymbol{Z}} \mid \boldsymbol{Z} \in \mathcal{L}(\mathcal{X})\}, \quad (3.7)$$

as the set of probability distribution functions corresponding to the random payoffs of the form $d + r\boldsymbol{X}$ that can be constructed out of the set $\mathcal{X}$. Note that $(\mathcal{F}(\mathcal{X}), \rho)$ constitutes a metric space (see also A.6). The modified continuity condition then reads as follows

(C5) $\succsim_i$ satisfies *weak continuity* if for all finite subsets $\mathcal{X} \subset L^1(\mathbb{R})$ it holds that the sets $\{F_{\boldsymbol{Z}} \in \mathcal{F}(\mathcal{X}) | \boldsymbol{Z} \succ_i \boldsymbol{Y}\}$ and $\{F_{\boldsymbol{Z}} \in \mathcal{F}(\mathcal{X}) | \boldsymbol{Z} \prec_i \boldsymbol{Y}\}$ are open in the metric space $(\mathcal{F}(\mathcal{X}), \rho)$ for all $\boldsymbol{Y} \in \mathcal{L}(\mathcal{X})$. [3] [4]

[3]If the preferences are complete, an equivalent statement is that $\{F_{\boldsymbol{Z}} \in \mathcal{F}(\mathcal{X}) | \boldsymbol{Z} \succsim_i \boldsymbol{Y}\}$ and $\{F_{\boldsymbol{Z}} \in \mathcal{F}(\mathcal{X}) | \boldsymbol{Z} \precsim_i \boldsymbol{Y}\}$ are closed sets in $(\mathcal{F}(\mathcal{X}), \rho)$ for all $\boldsymbol{Y} \in \mathcal{L}(\mathcal{X})$.

[4]For ease of notation, the sets $\{F_{\boldsymbol{Z}} \in \mathcal{F}(\mathcal{X}) | \boldsymbol{Z} \succsim_i \boldsymbol{Y}\}$ and $\{F_{\boldsymbol{Z}} \in \mathcal{F}(\mathcal{X}) | \boldsymbol{Z} \precsim_i \boldsymbol{Y}\}$ are often denoted by $\{\boldsymbol{Z} \in \mathcal{L}(\mathcal{X}) | \boldsymbol{Z} \succsim_i \boldsymbol{Y}\}$ and $\{\boldsymbol{Z} \in \mathcal{L}(\mathcal{X}) | \boldsymbol{Z} \precsim_i \boldsymbol{Y}\}$, respectively.

Note that since $\mathcal{V}(S)$ is finite by assumption, weak continuity implies that the preferences are also continuous on the set formulated in (3.5). Furthermore, note that weak continuity does not depend on a stochastic cooperative game $\Gamma \in SG(N)$.

Let $CG(N) \subset SG(N)$ denote the class of stochastic cooperative games for which $\mathcal{V}(S)$ is finite for each $S \subset N$ and the preferences $\succsim_i$ of each individual $i \in N$ satisfy the conditions (C1), (C2), (C3), and (C5).

3.2.2 Certainty Equivalents

Consider preferences $\{\succsim_i\}_{i \in N}$ such that for each $i \in N$ there exists a function $m_i : L^1(\mathbb{R}) \to \mathbb{R}$ satisfying

(M1) for all $\boldsymbol{X}, \boldsymbol{Y} \in L^1(\mathbb{R})$: $\boldsymbol{X} \succsim_i \boldsymbol{Y}$ if and only if $m_i(\boldsymbol{X}) \geq m_i(\boldsymbol{Y})$;

(M2) for all $\boldsymbol{X} \in L^1(\mathbb{R})$ and all $d \in \mathbb{R}$: $m_i(d + \boldsymbol{X}) = d + m_i(\boldsymbol{X})$.

The interpretation is that $m_i(\boldsymbol{X})$ equals the amount of money m for which agent i is indifferent between receiving the amount $m_i(\boldsymbol{X})$ with certainty and receiving the stochastic payoff $\boldsymbol{X}$. The amount $m_i(\boldsymbol{X})$ is called the certainty equivalent of $\boldsymbol{X}$. Condition (M1) states that agent i weakly prefers one stochastic payoff to another one if and only if the certainty equivalent of the former is greater than or equal to the certainty equivalent of the latter. Condition (M2) states that the certainty equivalent is linearly separable in the deterministic amount of money d.

Example 3.11 Consider the preferences based on a utility function of the form $U(t) = \beta e^{-\alpha t}$, $(t \in \mathbb{R})$, where $\beta < 0$ and $\alpha > 0$. The certainty equivalent of $\boldsymbol{X} \in L^1(\mathbb{R})$ can be defined by $m(\boldsymbol{X}) = U^{-1}(E(U(\boldsymbol{X})))$. It is easy to check that m satisfies condition (M1). For condition (M2), let $\boldsymbol{X} \in L^1(\mathbb{R})$ and $d \in \mathbb{R}$. Then $U^{-1}(t) = -\frac{1}{\alpha} \log\left(\frac{t}{\beta}\right)$ and

$$\begin{aligned} m(d + \boldsymbol{X}) &= U^{-1}(E(U(d + \boldsymbol{X}))) \\ &= -\tfrac{1}{\alpha} \log\left(\tfrac{1}{\beta} \int \beta e^{-\alpha(d+t)} dF_{\boldsymbol{X}}(t)\right) \\ &= -\tfrac{1}{\alpha} \log\left(e^{-\alpha d} \tfrac{1}{\beta} \int \beta e^{-\alpha t} dF_{\boldsymbol{X}}(t)\right) \\ &= d - \tfrac{1}{\alpha} \log\left(\tfrac{1}{\beta}(\int \beta e^{-\alpha t} dF_{\boldsymbol{X}}(t))\right) \\ &= d + m(\boldsymbol{X}). \end{aligned}$$

Finally, note that U is a monotonically increasing and concave function and thus implies risk averse behavior.

Example 3.12 Let the preferences $\succsim_\alpha$ be such that for $X, Y \in L^1(\mathbb{R})$ it holds that $\boldsymbol{X} \succsim_\alpha \boldsymbol{Y}$ if $\xi_\alpha(\boldsymbol{X}) \geq \xi_\alpha(\boldsymbol{Y})$, where $\alpha \in (0,1)$. With the certainty equivalent of $\boldsymbol{X} \in L^1(\mathbb{R})$ given by $m(\boldsymbol{X}) = \xi_\alpha(\boldsymbol{X})$, the conditions (M1) and (M2) are satisfied. That (M1) is fulfilled is straightforward. For (M2), note that

$$\begin{aligned}
\xi_\alpha(d+\boldsymbol{X}) &= \sup\{t \in \mathbb{R} \mid \mathbb{P}(\{d+\boldsymbol{X} < t\}) \leq \alpha\} \\
&= \sup\{t \in \mathbb{R} \mid \mathbb{P}(\{\boldsymbol{X} < t-d\}) \leq \alpha\} \\
&= \sup\{t+d \in \mathbb{R} \mid \mathbb{P}(\{\boldsymbol{X} < t\}) \leq \alpha\} \\
&= d + \sup\{t \in \mathbb{R} \mid \mathbb{P}(\{\boldsymbol{X} < t\}) \leq \alpha\} \\
&= d + \xi_\alpha(\boldsymbol{X}),
\end{aligned}$$

for all $d \in \mathbb{R}$ and all $\boldsymbol{X} \in L^1(\mathbb{R})$. Hence, $m(d+\boldsymbol{X}) = d + m(\boldsymbol{X})$.

Note that condition (M2) shows some resemblance with the transferable utility property defined in 2.3. Transferable utility implies that utility is linearly separable in the amount of money one receives while condition (M2) states that the certainty equivalent is linearly separable in the deterministic amount of money one receives. The similarities between (M2) and transferable utility go even further. As was the case for transferable utility, we can represent the stochastic benefits of each coalition by a single number if the preferences of each agent can be described by certainty equivalents. Consequently, we can represent a stochastic cooperative game by a TU-game.

Let $\Gamma \in SG(N)$ be a stochastic cooperative game satisfying conditions (M1) and (M2). Take $S \subset N$. An allocation $(d_i + r_i\boldsymbol{X})_{i\in S} \in \mathcal{Z}_\Gamma(S)$ is *Pareto optimal* for coalition S if there exists no allocation $(\hat{d}_i + \hat{r}_i\hat{\boldsymbol{X}})_{i\in S} \in \mathcal{Z}_\Gamma(S)$ such that $\hat{d}_i + \hat{r}_i\hat{\boldsymbol{X}} \succ_i d_i + r_i\boldsymbol{X}$ for all $i \in S$. Pareto optimal allocations are characterized by the following proposition, which is due to Suijs and Borm (1999).

Proposition 3.13 Let $\Gamma \in SG(N)$ satisfy conditions (M1) and (M2). Then $(d_i + r_i\boldsymbol{X})_{i\in S} \in \mathcal{Z}_\Gamma(S)$ is Pareto optimal if and only if

$$\begin{aligned}
\sum_{i\in S} m_i(d_i + r_i\boldsymbol{X}) &= \\
&\max\left\{\textstyle\sum_{i\in S} m_i(\hat{d}_i + \hat{r}_i\hat{\boldsymbol{X}}) \mid (\hat{d}_i + \hat{r}_i\hat{\boldsymbol{X}})_{i\in S} \in \mathcal{Z}_\Gamma(S)\right\}.
\end{aligned} \tag{3.8}$$

For interpreting condition (3.8), consider a particular allocation for coalition S and let each member pay the certainty equivalent of the random payoff he receives. Acting in this way, the initial wealth of each member does not change and, instead of the random payoff, the coalition now has to divide the certainty equivalents that have been paid by its members. Since the preferences are strictly increasing in the deterministic amount of money one receives, the more money a coalition can divide, the better it is for all its members. So, the best way to allocate the random benefits, is the one that maximizes the sum of the certainty equivalents. Furthermore, we can describe the random benefits of each coalition by the maximum sum of the certainty equivalents they can obtain, provided that this maximum exists, of course. This follows from the fact that for each $S \subset N$ it holds that

$$\begin{aligned} \{(m_i(d_i + r_i\boldsymbol{X}))_{i\in S} \mid (d_i + r_i\boldsymbol{X})_{i\in S} \in \mathcal{Z}_\Gamma(S)\} &= \\ \{x \in \mathbb{R}^S \mid \textstyle\sum_{i\in S} x_i \leq v_\Gamma(S)\}, \end{aligned} \tag{3.9}$$

where

$$v_\Gamma(S) = \max\left\{\sum_{i\in S} m_i(\hat{d}_i + \hat{r}_i\hat{\boldsymbol{X}}) \mid (\hat{d}_i + \hat{r}_i\hat{\boldsymbol{X}})_{i\in S} \in \mathcal{Z}_\Gamma(S)\right\}. \tag{3.10}$$

Expression (3.9) means that it does not matter for coalition S whether they allocate a random payoff $\boldsymbol{X} \in \mathcal{V}(S)$ or the deterministic amount $v_\Gamma(S)$. To see that this equality does indeed hold, note that the inclusion '$\subset$' follows immediately from the definition of $v_\Gamma(S)$. For the reverse inclusion '$\supset$', let $y \in \{x \in \mathbb{R}^S \mid \sum_{i\in S} x_i \leq v_\Gamma(S)\}$. Next, let $(d_i + r_i\boldsymbol{X})_{i\in S} \in \mathcal{Z}_\Gamma(S)$ be such that $\sum_{i\in S} m_i(d_i + r_i\boldsymbol{X}) = v_\Gamma(S)$ and define $\delta_i = y_i + d_i - m_i(d_i + r_i\boldsymbol{X})$ for each $i \in S$. Since $\sum_{i\in S} \delta_i \leq 0$ and

$$\begin{aligned} m_i(\delta_i + r_i\boldsymbol{X}) &= m_i(\delta_i - d_i + d_i + r_i\boldsymbol{X}) \\ &= \delta_i - d_i + m_i(d_i + r_i\boldsymbol{X}) \\ &= y_i + d_i - m_i(d_i + r_i\boldsymbol{X}) - d_i + m_i(d_i + r_i\boldsymbol{X}) = y_i \end{aligned}$$

for all $i \in S$ it holds that $y \in \{(m_i(d_i + r_i\boldsymbol{X}))_{i\in S} \mid (d_i + r_i\boldsymbol{X})_{i\in S} \in \mathcal{Z}_\Gamma(S)\}$. Hence, the benefits of coalition S can be represented by $v_\Gamma(S)$. So, if for a stochastic cooperative game $\Gamma \in SG(N)$ the value $v_\Gamma(S)$ is well defined for each coalition $S \subset N$ we can also describe the game Γ by a TU-game (N, v_Γ) with $v_\Gamma(S)$ as in (3.10). Let $MG(N) \subset SG(N)$ denote the class of stochastic cooperative games for which the conditions (M1) and (M2) are satisfied and the game (N, v_Γ) is well defined.

4 The Core, Superadditivity, and Convexity

This chapter recapitulates the main results of Suijs and Borm (1999). They introduce the core for stochastic cooperative games, which, given its interpretation for both TU- and NTU-games, extends fairly straightforward to the class of stochastic cooperative games. A balancedness condition like the one for TU-games, however, does not yet exist. They only provide such a condition for the subclass MG(N) of stochastic cooperative games for which certainty equivalents are well defined. Besides the core, they also extend the definitions of superadditivity and convexity to the class of stochastic cooperative games. Suijs and Borm (1999) show that a convex stochastic cooperative game is superadditive and has a nonempty core. Furthermore, they define marginal vectors and show that for each convex stochastic cooperative game all marginal vectors belong to the core.

4.1 The Core of a Stochastic Cooperative Game

Section 2.3 and Section 2.4 considered the core for TU-games and NTU-games, respectively. For these games, an allocation is a core-allocation if no coalition has an incentive to part company with the grand coalition. This same principle is used to define a core for stochastic cooperative games. Therefore, we need to specify under which conditions a coalition has an incentive to leave the grand coalition. So, let $\Gamma \in SG(N)$ be a stochastic cooperative game and let $(d_i + r_i\boldsymbol{X})_{i\in N} \in \mathcal{Z}_\Gamma(N)$ be an allocation for N. Coalition S has an incentive to leave coalition N and start cooperating on its own if it can improve on the payoff of each of its members. This means that there exists an allocation $(\hat{d}_i + \hat{r}_i\hat{\boldsymbol{X}})_{i\in S} \in \mathcal{Z}_\Gamma(S)$ such that $\hat{d}_i + \hat{r}_i\hat{\boldsymbol{X}} \succ_i d_i + r_i\boldsymbol{X}$ for all $i \in S$. The *core* of a stochastic cooperative game $\Gamma \in SG(N)$ is thus defined by

$$\begin{aligned} \mathcal{C}(\Gamma) = \{&(d_i + r_i\boldsymbol{X})_{i\in N} \in \mathcal{Z}_\Gamma(N)| \\ &\forall_{S\subset N}\ \nexists_{(\hat{d}_i+\hat{r}_i\hat{\boldsymbol{X}})_{i\in S}\in\mathcal{Z}_\Gamma(S)} \forall_{i\in S} : \hat{d}_i + \hat{r}_i\hat{\boldsymbol{X}} \succ_i d_i + r_i\boldsymbol{X}\}. \end{aligned} \tag{4.1}$$

A stochastic cooperative game $\Gamma \in SG(N)$ with a nonempty core is called *balanced*. Furthermore, if the core of every subgame $\Gamma_{|S}$ is nonempty, then Γ is called *totally balanced*. The subgame $\Gamma_{|S}$ is given by $(S, \mathcal{V}_{|S}, \{\succsim_i\}_{i\in S})$ where $\mathcal{V}_{|S}(T) = \mathcal{V}(T)$ for all $T \subset S$.

For the class of TU- and NTU-games we saw that the core can be an empty set. The same holds for the class of stochastic cooperative games. We will provide an example of a game with an empty core later in this section, but first we take a closer look at the class $MG(N)$. Recall that for each $\Gamma \in MG(N)$ there corresponds a TU-game (N, v_Γ) with

$$v_\Gamma(S) = \max\{\sum_{i\in S} m_i(d_i + r_i\boldsymbol{X}) | \, (d_i + r_i\boldsymbol{X})_{i\in S} \in \mathcal{Z}_\Gamma(S)\},$$

for all $S \subset N$.

Theorem 4.1 Let $\Gamma \in MG(N)$. Then $\mathcal{C}(\Gamma) = \emptyset$ if and only if $C(v_\Gamma) = \emptyset$.

Formulated in terms of allocations this theorem reads as follows: Let $\Gamma \in MG(N)$. If $(d_i + r_i\boldsymbol{X})_{i\in N} \in \mathcal{Z}_\Gamma(N)$ and $x \in \mathbb{R}^N$ are such that $m_i(d_i + r_i\boldsymbol{X}) = x_i$ for all $i \in N$ then

$$(d_i + r_i\boldsymbol{X})_{i\in N} \in \mathcal{C}(\Gamma) \text{ if and only if } x \in C(v_\Gamma).$$

Furthermore, we know that the core of the game (N, v_Γ) is nonempty if and only if the balancedness condition formulated in (2.10) is satisfied. Theorem 4.1 then provides an easy way to construct a stochastic cooperative game with an empty core.

Example 4.2 Consider a three-person stochastic cooperative game $\Gamma \in MG(N)$ with $\mathcal{V}(\{i\}) = \{0\}$, $i = 1, 2, 3$ and $\mathcal{V}(S) = \{\boldsymbol{X}_S\}$ with $\boldsymbol{X}_S \sim U(0, 2)$ if $\#S \geq 2$. Furthermore, let all agents be risk neutral. This means that agent i's certainty equivalent of a random payoff $\boldsymbol{X} \in L^1(\mathbb{R})$ equals $m_i(\boldsymbol{X}) = E(\boldsymbol{X})$. Consequently, we have for $(d_i + r_i\boldsymbol{X})_{i\in S} \in \mathcal{Z}_\Gamma(S)$ that

$$\begin{aligned}\sum_{i\in S} m_i(d_i + r_i\boldsymbol{X}) &= \sum_{i\in S} E(d_i + r_i\boldsymbol{X}) = \sum_{i\in S} d_i + E\left(\sum_{i\in S} r_i\boldsymbol{X}\right) \\ &= \sum_{i\in S} d_i + E(\boldsymbol{X}).\end{aligned}$$

The corresponding TU-game (N, v_Γ) then equals $v_\Gamma(S) = 0$ if $\#S = 1$ and $v_\Gamma(S) = 1$ if $\#S \geq 2$. For the game (N, v_Γ) it holds that $C(v_\Gamma) = \emptyset$. Hence, by Theorem 4.1, $\mathcal{C}(\Gamma) = \emptyset$.

So, for the class $MG(N)$ of stochastic cooperative games we can relatively easy check whether the core is empty or not. For the more general class $SG(N)$, however, necessary *and* sufficient conditions for nonemptiness of the core do not yet exist.

4.2 Superadditive Games

For introducing superadditivity for stochastic cooperative games recall that for both TU- and NTU-games the underlying idea of superadditivity is that two disjoint coalitions can do (weakly) better by forming one coalition. Therefore, we propose the following definition of superadditivity, which conceptually is not only applicable to stochastic cooperative games, but also to TU- and NTU-games. Let $\Gamma \in SG(N)$ and define for each $S \subset N$ the set of individually rational allocations by

$$\mathcal{IR}_\Gamma(S) = \{(d_i+r_i\boldsymbol{X})_{i\in S} \in \mathcal{Z}_\Gamma(S) | \forall_{i\in S} \nexists_{\boldsymbol{Y}\in\mathcal{V}(\{i\})} : \boldsymbol{Y} \succ_i d_i+r_i\boldsymbol{X}\}. \quad (4.2)$$

Then Γ is called *superadditive* if for any disjoint $S, T \subset N$ it holds that for every $(d_i^S + r_i^S\boldsymbol{X}^S)_{i\in S} \in \mathcal{IR}_\Gamma(S)$ and every $(d_i^T + r_i^T\boldsymbol{X}^T)_{i\in T} \in \mathcal{IR}_\Gamma(T)$ there exists an allocation $(d_i^{S\cup T} + r_i^{S\cup T}\boldsymbol{X}^{S\cup T})_{i\in S\cup T} \in \mathcal{Z}_\Gamma(S \cup T)$ such that

$$\begin{aligned} d_i^{S\cup T} + r_i^{S\cup T}\boldsymbol{X}^{S\cup T} &\succsim_i d_i^S + r_i^S\boldsymbol{X}^S \quad \text{for all } i \in S, \\ d_i^{S\cup T} + r_i^{S\cup T}\boldsymbol{X}^{S\cup T} &\succsim_i d_i^T + r_i^T\boldsymbol{X}^T \quad \text{for all } i \in T. \end{aligned} \quad (4.3)$$

So whatever allocation the coalitions S and T agree on separately, they can always (weakly) improve their payoffs by forming one large coalition. Formulated in the context of NTU-games this definition reads as follows. For all disjoint $S, T \subset N$ it holds that for each allocation $x^S \in V(S)$ and each allocation $x^T \in V(T)$ there exists an allocation $x^{S\cup T} \in V(S \cup T)$ such that $x_i^{S\cup T} \geq x_i^S$ for all $i \in S$ and $x_i^{S\cup T} \geq x_i^T$ for all $i \in T$. Then it is not difficult to check that this definition is equivalent to the definition of superadditivity for NTU-games given in (2.18), i.e., for all disjoint $S, T \subset N$ it holds that $V(S) \times V(T) \subset V(S \cup T)$.

Theorem 4.3 Let $\Gamma \in MG(N)$. Then Γ is superadditive if and only if (N, v_Γ) is superadditive.

Note that the game in Example 4.2 is superadditive. So, similar as for TU- and NTU-games, we need a stronger condition for nonemptiness of the core than superadditivity. One such condition is convexity.

4.3 Convex Games

For explaining the definition of convexity, let us take convexity for TU-games as formulated in (2.6) as a starting point. A TU-game (N, v) is called convex if for each $U \subset N$ and each $S \subset T \subset N\backslash U$ it holds that

$$v(S \cup U) - v(S) \leq v(T \cup U) - v(T). \quad (4.4)$$

This means that for a coalition it is more profitable to join a larger coalition. Now, we apply this idea to stochastic cooperative games. Let $\Gamma \in SG(N)$. Then Γ is called *convex* if for each $U \subset N$ and each $S \subset T \subset N \backslash U$ the following statement is true: for all $(d_i^S + r_i^S \boldsymbol{X}^S)_{i \in S} \in \mathcal{IR}_\Gamma(S)$, all $(d_i^T + r_i^T \boldsymbol{X}^T)_{i \in T} \in \mathcal{IR}_\Gamma(T)$ and all $(d_i^{S \cup U} + r_i^{S \cup U} \boldsymbol{X}^{S \cup U})_{i \in S \cup U} \in \mathcal{Z}_\Gamma(S \cup U)$ satisfying

$$d_i^{S \cup U} + r_i^{S \cup U} \boldsymbol{X}^{S \cup U} \succsim_i d_i^S + r_i^S \boldsymbol{X}^S,$$

for all $i \in S$, there exists an allocation $(d_i^{T \cup U} + r_i^{T \cup U} \boldsymbol{X}^{T \cup U})_{i \in T \cup U} \in \mathcal{Z}_\Gamma(T \cup U)$ such that

$$\begin{aligned} d_i^{T \cup U} + r_i^{T \cup U} \boldsymbol{X}^{T \cup U} &\succsim_i d_i^T + r_i^T \boldsymbol{X}^T && \text{for all } i \in T, \text{ and} \\ d_i^{T \cup U} + r_i^{T \cup U} \boldsymbol{X}^{T \cup U} &\succsim_i d_i^{S \cup U} + r_i^{S \cup U} \boldsymbol{X}^{S \cup U} && \text{for all } i \in U. \end{aligned} \tag{4.5}$$

So whatever individually rational allocation the coalitions S and T agree on separately, then given an allocation for coalition $S \cup U$ such that coalition S is willing to let coalition U join, the members of coalition U can obtain (weakly) better payoffs by joining the larger coalition T. By taking $S = \emptyset$ in the definition of convexity it follows immediately that convex games are superadditive.

For the next theorem, we need to define marginal vectors for stochastic cooperative games. Recall that a marginal vector with respect to an order $\sigma \in \Pi_N$ gives each agent his marginal contribution to the benefits of the coalition that is formed by the preceding agents. Keeping this in mind we can define a marginal vector. For this purpose, take $\Gamma \in CG(N)$ and let us abbreviate the notation $(d_i + r_i \boldsymbol{X})_{i \in S}$ of an allocation to $\boldsymbol{Y}$. Take $i \in N$, $S \subset N \backslash \{i\}$ with $S \neq \emptyset$ and $\boldsymbol{Y} \in \mathcal{Z}_\Gamma(S)$. Define

$$\mathcal{B}(S, \boldsymbol{Y}, i) = \{\hat{\boldsymbol{Y}} \in \mathcal{Z}_\Gamma(S \cup \{i\}) | \forall_{j \in S} : \hat{\boldsymbol{Y}}_j \succsim_j \boldsymbol{Y}_j\},$$

as the set of all allocations for coalition $S \cup \{i\}$ which all members of S weakly prefer to the allocation $\boldsymbol{Y} \in \mathcal{Z}_\Gamma(S)$. Note that since there is no lower bound on agent i's payoff, the set $\mathcal{B}(S, \boldsymbol{Y}, i)$ is nonempty. Furthermore, let $\boldsymbol{B}(S, \boldsymbol{Y}, i) \in \mathcal{B}(S, \boldsymbol{Y}, i)$ be the most preferred allocation for agent i in this set, that is, $\boldsymbol{B}_i(S, \boldsymbol{Y}, i) \succsim_i \hat{\boldsymbol{Y}}_i$ for all $\hat{\boldsymbol{Y}} \in \mathcal{B}(S, \boldsymbol{Y}, i)$. Since the preferences are assumed to be complete, transitive and weakly continuous such a best allocation exists. A marginal vector $\boldsymbol{M}^\sigma(\mathcal{V}) \in \mathcal{Z}_\Gamma(N)$ with respect to order $\sigma \in \Pi_N$ is now constructed as follows. Let $i_k \in N$ be such that $\sigma(i_k) = k$ for $k = 1, 2, \ldots, n$. Let $\boldsymbol{Y}^1 \in \mathcal{Z}_\Gamma(\{i_1\})$ be individually rational. Thus $\boldsymbol{Y}^1 \in \mathcal{IR}_\Gamma(\{i_1\})$. Take $\boldsymbol{Y}^2 \in \mathcal{Z}_\Gamma(\{i_1, i_2\})$ such that $\boldsymbol{Y}^2 = \boldsymbol{B}(\{i_1\}, \boldsymbol{Y}^1, i_2)$. Next, take $\boldsymbol{Y}^3 \in \mathcal{Z}_\Gamma(\{i_1, i_2, i_3\})$ such that $\boldsymbol{Y}^3 = \boldsymbol{B}(\{i_1, i_2\}, \boldsymbol{Y}^2, i_3)$. Continuing this procedure yields an allocation $\boldsymbol{Y}^n \in \mathcal{Z}_\Gamma(N)$. Finally, define a *marginal vector* with respect to order σ by $\boldsymbol{M}^\sigma(\mathcal{V}) = \boldsymbol{Y}^n$.

Theorem 4.4 Let $\Gamma \in CG(N)$ be a stochastic cooperative game. If Γ is convex, then all marginal vectors $\boldsymbol{M}^\sigma(\mathcal{V})$, $\sigma \in \Pi_N$ belong to the core. Hence, $\mathcal{C}(\Gamma) \neq \emptyset$.

Theorem 4.4 also holds for NTU-games if we formulate convexity based on (4.5) in the following way. An NTU-game (N, V) satisfies convexity if for every $U \subset N$ and every $S \subset T \subset N \backslash U$ the following statement holds. For every individually rational $x^S \in V(S)$, every individually rational $x^T \in V(T)$ and every $x^{S \cup U} \in V(S \cup U)$ such that $x_i^{S \cup U} \geq x_i^S$ for all $i \in S$, there exists $x^{T \cup U} \in V(T \cup U)$ satisfying

$$\begin{aligned} x_i^{T \cup U} &\geq x_i^T && \text{for all } i \in T, \text{ and} \\ x_i^{T \cup U} &\geq x_i^{S \cup U} && \text{for all } i \in U. \end{aligned}$$

In this way, convex NTU-games are totally balanced and every marginal vector belongs to the core. This also implies that our definition of convexity is not equivalent to ordinal or cardinal convexity - see Vilkov (1977) and Sharkey (1981) for their respective definitions - since for both notions of convexity not all marginal vectors belong to the core. Furthermore, for NTU-games we can show that the reverse of Theorem 4.4 does not hold, that is an NTU-game need not be convex if all marginal vectors belong to the core. A counterexample is provided in Figure 4.3. As one can see, all marginal vectors are core-allocations. Convexity, however, is not satisfied in this case for the following reason. Take $S = \{3\}$, $T = \{2, 3\}$, and $U = \{1\}$. Since 0 is an individually rational allocation for coalition S, the allocation x for coalition $S \cup U$ satisfies the condition that it improves the payoffs of the members of coalition S, i.e. agent 3. Now, if the game is convex, there must exist an allocation for $T \cup U = \{1, 2, 3\}$ that is (weakly) better than the allocation y for the individuals in T, and (weakly) better than x for the individuals in U. But as Figure 4.3 shows, such an allocation does not exist.

Finally, let us focus once more on the class $MG(N)$ of stochastic cooperative games.

Theorem 4.5 Let $\Gamma \in MG(N)$. Then Γ is convex if and only if (N, v_Γ) is convex.

In a similar way as for Theorem 4.5 one can show that in the context of TU-games the definitions of convexity provided in (2.6) and the TU-formulation of (4.5) are equivalent. Hence, the convexity introduced in this section is an extension of the convexity definition for TU-games.

4.4 Remarks

When introducing convexity for stochastic cooperative games we also defined marginal vectors for these type of games. This naturally raises the question if we

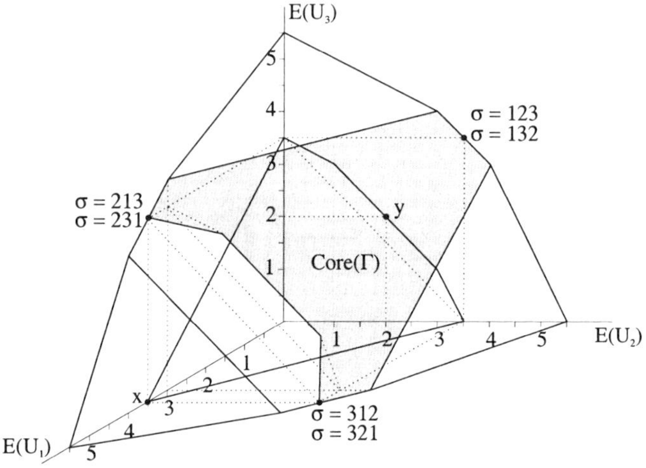

Figure 4.1

can also define a Shapley value for stochastic cooperative games. For TU-games, the Shapley value is defined as the average of all marginal vectors. For stochastic cooperative games though, taking the average of the marginal vectors does not give the desired result. As is the case for NTU-games, the average need no longer be Pareto optimal. Furthermore, for each order $\sigma \in \Pi_N$ the marginal vector is not necessarily uniquely determined. Although the expected utility levels are uniquely determined, this need not be the case for the random payoffs itself. Hence, for each order there can exist several, maybe even an infinite number of marginal vectors.

So, when extending the Shapley value to stochastic cooperative games one encounters two major problems. First, the average of the marginal vectors can violatc Pareto optimality and, second, a marginal vector need not be unique.

5 Nucleoli for Stochastic Cooperative Games

The nucleolus, a solution concept for TU-games, originates from Schmeidler (1969). This solution concept yields an allocation that minimizes the excesses of the coalitions in a lexicographical way. The excess describes how dissatisfied a coalition is with the proposed allocation. The larger the excess of a particular allocation, the more a coalition is dissatisfied with this allocation. For Schmeidler's nucleolus the excess is defined as the difference between the payoff a coalition can obtain when cooperating on its own and the payoff received by the proposed allocation. So, when less is allocated to a coalition, the excess of this coalition increases and the other way round.

Since the nucleolus depends mainly on the definition of the excess, other nucleoli are found when different definitions of excesses are used. Such a general approach can be found in Potters and Tijs (1992) and Maschler, Potters and Tijs (1992). They introduced the general nucleolus as the solution that minimizes the maximal excess of the coalitions, using generally defined excess functions.

A similar argument holds for stochastic cooperative games. If we can specify the excesses we can define a nucleolus for these games. Unfortunately, this is not that simple. Defining excess functions for stochastic cooperative games appears to be not as straightforward as for deterministic cooperative games. Indeed, how should one quantify the difference between the random payoff a coalition can achieve on its own and the random payoff received by the proposed allocation when the behavior towards risk can differ between the members of this coalition? Furthermore, the excess of one coalition should be comparable to the excess of another coalition.

Charnes and Granot (1976) introduced a nucleolus for chance-constrained games. There, the excess was based on the probability that the payoff a coalition can obtain when cooperating on its own, exceeds the payoff they receive in the proposed allocation. Indeed, it is quite reasonable to assume that a coalition is less satisfied with the proposed allocation if this probability increases. The resulting nucleolus, however, is not an extension of Schmeidler's nucleolus, for if the payoffs to coalitions are deterministic, the probabilities that describe the excesses equal either 0 or 1.

This chapter introduces two different nucleoli, both of them being extensions of the nucleolus for TU-games. The corresponding excess functions are based on the following interpretation of the excess function as defined in Schmeidler (1969). Bearing the conditions of the core in mind, this excess can be interpreted as follows. Given an allocation of the grand coalition's payoff we distinguish two cases. In the first case, a coalition wants to leave the grand coalition. Then the excess equals the minimal amount of money a coalition needs on top of what they already receive such that this coalition is willing to stay in the grand coalition. In the second case, a coalition has no incentive to leave the grand coalition. Then the excess equals minus the maximal amount of money that can be taken away from this coalition such that this coalition still has no incentive to leave the grand coalition.

Reasoning in this way, one obtains different excess functions, depending on which payoff is affected by the reshuffling of the money. For the first excess function that we introduce, money is added to or subtracted from the payoff $\boldsymbol{X}_S$ that a coalition S can obtain, while for the other excess function, money is added to or subtracted from the payoff that is proposed by the grand coalition.

In the remainder of this chapter, we assume for ease of notation that $\mathcal{V}(S) = \{\boldsymbol{X}_S\}$ for all $S \in N$. The results presented in this chapter also hold true though, if $\mathcal{V}(S)$ is finite for all $S \subset N$. Furthermore, we need some additional notation. Let

$$I_\Gamma(S) = \{(d,r) \in \mathbb{R}^S \times \Delta^S | \forall_{i\in S} : d_i + r_i \boldsymbol{X}_S \succsim_i \boldsymbol{X}_{\{i\}}\},$$

denote the set of possibly nonfeasible individually rational allocations for coalition S. Note that $I_\Gamma(S)$ is a subset of $\mathbb{R}^S \times \Delta^S$ and not a subset of $L^1(\mathbb{R})^S$. So it contains the pairs (d,r) that correspond to an allocation $(d_i + r_i \boldsymbol{X}_S)_{i\in S}$ instead of the allocations itself. In this chapter we refer to both (d,r) and $(d_i + r_i \boldsymbol{X}_S)_{i\in S}$ as an allocation. An allocation $(d,r) \in I_\Gamma(S)$ is called feasible if $\sum_{i\in S} d_i \leq 0$. So, let

$$IR_\Gamma(S) = \{(d,r) \in I_\Gamma(S) | \sum_{i\in S} d_i \leq 0\},$$

denote the set of feasible individually rational allocations for coalition S. We assume that $IR_\Gamma(S) \neq \emptyset$ for all $S \subset N$. Note that this assumption is satisfied if Γ is superadditive. Moreover, it should be noted that a coalition S is unlikely to be formed whenever $IR_\Gamma(S) = \emptyset$. Since in that case there is for every allocation of $\boldsymbol{X}_S$ at least one member of S whose payoff is not individually rational. Hence, he would be better off by leaving coalition S and forming a coalition on his own. Finally, let

$$\begin{aligned} PO_\Gamma(S) = \ & \{(d,r) \in IR_\Gamma(S) | \\ & \nexists_{(d',r') \in IR_\Gamma(S)} \forall_{i\in S} : d'_i + r'_i \boldsymbol{X}_S \succ_i d_i + r_i \boldsymbol{X}_S\}, \end{aligned}$$

denote the set of feasible Pareto optimal allocations for S. Note that assumption (C2) in Section 3.2.1 implies that $\sum_{i\in S} d_i = 0$ whenever $(d,r) \in PO_\Gamma(S)$.

The following propositions show that the sets $IR_\Gamma(S)$ and $PO_\Gamma(S)$ are compact subsets for the class $CG(N)$ of stochastic cooperative games. The proofs can be found in Appendix 5.4.

Proposition 5.1 The set of individually rational allocations $IR_\Gamma(S)$ is a compact subset of $I_\Gamma(S)$ for each coalition $S \subset N$.

Proposition 5.2 The set of Pareto optimal allocations $PO_\Gamma(S)$ is a compact subset of $I_\Gamma(S)$ for each coalition $S \subset N$.

5.1 Nucleolus $\mathcal{N}^1$

For defining a nucleolus one needs to specify for each coalition $S \subset N$ an excess function. The excess function assigns to each allocation $(d_i + r_i \boldsymbol{X}_N)_{i\in N} \in \mathcal{Z}_\Gamma(N)$ of the grand coalition N a real number representing the complaint of coalition S. The larger the complaint of a coalition the more this coalition is dissatisfied with the proposed allocation. One way to define the excess function $E_S^1 : IR_\Gamma(N) \to \mathbb{R}$ is the following

$$E_S^1(d,r) = -\min_{(d',r')\in IR_\Gamma(S)} \left\{ \sum_{i\in S} \delta_i | \forall_{i\in S} : d'_i + \delta_i + r'_i \boldsymbol{X}_S \sim_i d_i + r_i \boldsymbol{X}_N \right\}. \quad (5.1)$$

For an interpretation of the excess function, let us focus on the core conditions. So, given a proposed allocation $(d,r) \in IR_\Gamma(N)$, does coalition S have an incentive to part company with the grand coalition or not? For this purpose, we first determine when a coalition is indifferent between staying in the grand coalition and leaving. Given an allocation $(d,r) \in IR_\Gamma(N)$ for the grand coalition, coalition S is indifferent between staying and leaving, if there exists a Pareto optimal allocation $(d',r') \in PO_\Gamma(S)$ such that each agent $i \in S$ is indifferent between receiving the payoff $d'_i + r'_i \boldsymbol{X}_S$ and the payoff $d_i + r_i \boldsymbol{X}_N$. Thus, coalition S cannot do strictly better by leaving the grand coalition, but if they do split off they can allocate their random payoff in such a way that no member is strictly worse off.

Now, suppose that coalition S does have an incentive to leave the grand coalition. Then there exists an allocation $(d',r') \in PO_\Gamma(S)$ such that each agent $i \in S$ strictly prefers the payoff $d'_i + r'_i \boldsymbol{X}_S$ to $d_i + r_i \boldsymbol{X}_N$. This means that coalition S can obtain too much on its own compared to what they receive in the allocation $(d_i + r_i \boldsymbol{X}_N)_{i\in N}$. The excess then expresses the maximum deterministic amount of money $\sum_{i\in S} \delta_i$, with which the random payoff of coalition S can decrease, so that they are indifferent between staying in the grand coalition and leaving.

For a graphical illustration of how this excess function works, consider Figure 5.1, which depicts the expected utility levels that the coalition $S = \{i, j\}$ can obtain in some stochastic cooperative game $\Gamma \in CG(N)$.[1] Here, the grand coalition proposes the allocation $(d, r) \in IR_\Gamma(N)$. Since the allocation (d, r) falls in the interior of $IR_\Gamma(S)$, coalition S has an incentive to leave the grand coalition. By taking away an additional amount of money x from coalition S, the set of attainable allocations becomes

$$A(x) = \{(d' + \delta, r') \in IS_\Gamma(S) |\ (d', r') \in IR_S(\Gamma) \text{and} \quad \exists_{\delta \in \mathbb{R}^S} : \textstyle\sum_{i \in S} \delta_i \leq x\}. \tag{5.2}$$

We reduce the set of attainable allocations until the allocation $(d, r) \in IR_\Gamma(N)$ hits the boundary of $A(x)$, which is the case when $x = E^1_S(d, r)$.

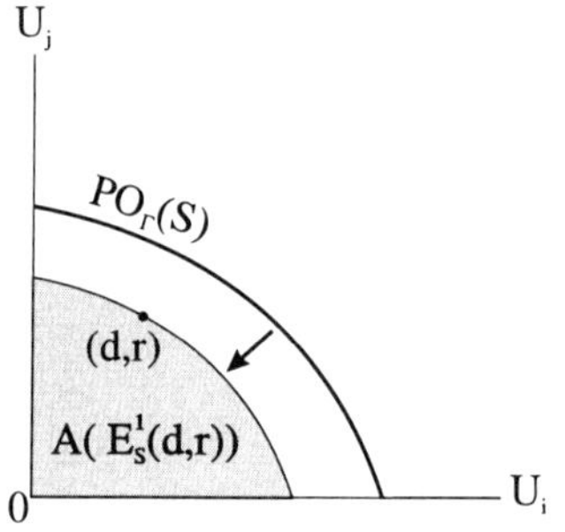

Figure 5.1

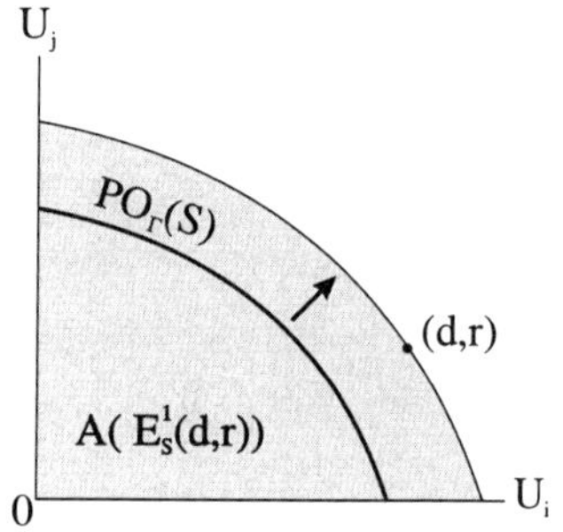

Figure 5.2

The excess is determined in a similar way when coalition S does not have an incentive to part company with the grand coalition N. Then there exists no allocation $(d', r') \in PO_\Gamma(S)$ such that each agent $i \in S$ strictly prefers the payoff $d'_i + r'_i \boldsymbol{X}_S$ to the payoff $d_i + r_i \boldsymbol{X}_N$. In other words, coalition S can obtain too little on its own compared to what they receive in the allocation $(d_i + r_i \boldsymbol{X}_N)_{i \in N}$. The excess then expresses the minimum deterministic amount of money, with which the random payoff of coalition S should increase, so that they are just willing to stay in the grand coalition, that is for any amount higher than $\sum_{i \in S} \delta_i$, coalition S has an incentive to leave the grand coalition.

This situation is illustrated in Figure 5.2. Note that since the allocation $(d, r) \in IR_\Gamma(N)$ falls outside $IR_\Gamma(S)$, coalition S does indeed have no incentive to leave

[1] Note that Figure 5.2 and the forthcoming figures in this chapter do not arise from a concrete stochastic cooperative game.

the grand coalition. In this case, we enlarge the set of attainable allocations by giving some money x until (d,r) hits the boundary of the set $A(x)$. Again, this happens when $x = E^1_S(d,r)$.

Summarizing, the excess equals the amount of money $\sum_{i\in S}\delta_i \in \mathbb{R}$ for which coalition S is indifferent between staying in and leaving the grand coalition, when they have a random payoff $\boldsymbol{X}_S - \sum_{i\in S}\delta_i$ to allocate. Moreover, if $(d_i + r_i\boldsymbol{X}_N)_{i\in N}$ and $(d'_i + r'_i\boldsymbol{X}_N)_{i\in N}$ are allocations of X_N such that each individual $i \in S$ prefers $d_i + r_i\boldsymbol{X}_N$ to $d'_i + r'_i\boldsymbol{X}_N$ then $E^1_S(d,r) < E^1_S(d',r')$. Hence, the excess decreases when each agent $i \in S$ improves his payoff. So, in a specific way the excess $E_S(d,r)$ describes how much coalition S is satisfied with the allocation $(d_i + r_i\boldsymbol{X}_N)_{i\in N}$. Furthermore, since all agents' preferences are monotonically increasing in the amount of money d they receive (see assumption (C2)), it is reasonable to say that one coalition is more satisfied with a particular allocation than another coalition if the first coalition needs less money to be satisfied than the latter one, or, in other words, if the excess of the first coalition is less than the excess of the latter. This last observation leads to the following definition of a nucleolus.

Let $\Gamma \in CG(N)$ be a cooperative game with stochastic payoffs and let

$$E^1_S(d,r) = -\min_{(d',r')\in IR_\Gamma(S)}\{\sum_{i\in S}\delta_i | \forall_{i\in S} : d'_i + \delta_i + r'_i\boldsymbol{X}_S \sim_i d_i + r_i\boldsymbol{X}_N\} \tag{5.1}$$

describe the excess of coalition S at allocation $(d,r) \in IR_\Gamma(N)$. Next, denote by $E^1(d,r)$ the vector of excesses at allocation $(d_i + r_i\boldsymbol{X}_N)_{i\in N}$ and let $\theta \circ E^1(d,r)$ denote the vector of excesses with its elements arranged in decreasing order. The *nucleolus* $\mathcal{N}^1(\Gamma)$ of the game $\Gamma \in CG(N)$ is then defined by

$$\begin{aligned}\mathcal{N}^1(\Gamma) = \{(d,r) \in IR_\Gamma(N)|\ \forall_{(d',r')\in IR_\Gamma(N)} :& \\ \theta \circ E^1(d,r) \leq_{lex} \theta \circ E^1(d',r')\},&\end{aligned} \tag{5.3}$$

where $\leq_{lex}$ is the lexicographical ordering. The following theorem shows that this nucleolus is a well defined solution concept for the class $CG(N)$ of stochastic cooperative games.

Theorem 5.3 Let $\Gamma \in CG(N)$ be a stochastic cooperative game. If $IR_\Gamma(N) \neq \emptyset$ then $\mathcal{N}^1(\Gamma) \neq \emptyset$.

PROOF: In proving the nonemptiness of the nucleolus $\mathcal{N}^1(\Gamma)$ we make use of the results stated in Maschler et al. (1992). They introduced a nucleolus for a more general framework and showed that the nucleolus is nonempty if the domain is compact and the excess functions are continuous. Thus, it suffices to show that $IR_\Gamma(N)$ is compact and that $E^1_S((d,r))$ is continuous in (d,r) for each $(d,r) \in IR_\Gamma(N)$ and each $S \subset N$.

The compactness of $IR_\Gamma(N)$ follows from Proposition 5.1. Next, we show that $E_S^1(d,r)$ is continuous in (d,r) for each $S \subset N$. Therefore, take $S \subset N$. For each $i \in S$ define the function $f_i : \mathbb{R}^N \times \Delta^N \times \Delta^S \to \mathbb{R}$ such that $f_i(d,r,r') + r'_i \boldsymbol{X}_S \sim_i d_i + r_i \boldsymbol{X}_N$. Then

$$\begin{aligned} E_S^1(d,r) &= -\min_{(d',r') \in IR_\Gamma(S)} \{\sum_{i \in S} \delta_i | \forall_{i \in S} : d'_i + \delta_i + r'_i \boldsymbol{X}_S \sim_i d_i + r_i \boldsymbol{X}_N\} \\ &= -\min_{(d',r') \in IR_\Gamma(S)} \{\sum_{i \in S} \delta_i | \forall_{i \in S} : \delta_i = f_i(d,r,r') - d'_i\} \\ &= -\min_{(d',r') \in IR_\Gamma(S)} \left(\sum_{i \in S} f_i(d,r,r') - \sum_{i \in S} d'_i\right) \\ &= -\min_{(d',r') \in IR_\Gamma(S)} \sum_{i \in S} f_i(d,r,r'). \end{aligned}$$

Hence, it is sufficient to show that f_i is a continuous function for each $i \in S$. We prove this in two steps.

First, we show that f_i is indeed a function. Given $(d,r,r') \in \mathbb{R}^N \times \Delta^N \times \Delta^S$, define

$$\begin{aligned} T_- &= \{t \in \mathbb{R} | d_i + r_i \boldsymbol{X}_N \succsim_i t + r'_i \boldsymbol{X}_S\}, \text{ and} \\ T_+ &= \{t \in \mathbb{R} | d_i + r_i \boldsymbol{X}_N \precsim_i t + r'_i \boldsymbol{X}_S\}. \end{aligned}$$

By condition (C3) both sets are nonempty. Condition (C2) implies that T_- is an interval unbounded from the left, and T_+ is an interval unbounded from the right. By continuity, both intervals are closed. Since $\succsim_i$ is complete, $T_- \cup T_+ = \mathbb{R}$. Hence, $T_- \cap T_+ \neq \emptyset$. Again, by condition (C2), their intersection, which equals f_i, is a singleton.

Second, we prove that f_i is continuous for each $i \in S$. Take $i \in S$ and $(d,r,r') \in \mathbb{R}^N \times \Delta^N \times \Delta^S$. Let $(d^k, r^k, r'^k)_{k \in \mathbb{N}}$ be a sequence in $\mathbb{R}^N \times \Delta^N \times \Delta^S$ converging to (d,r,r'). Take $\varepsilon > 0$ and define $\mathcal{X} = \{\boldsymbol{X}_S | S \subset N\}$, $\mathcal{L}(\mathcal{X})$ as in (3.6), and

$$O^\varepsilon = \{\boldsymbol{X} \in \mathcal{L}(\mathcal{X}) | d_i + r_i \boldsymbol{X}_N - \varepsilon \prec_i \boldsymbol{X} \prec_i d_i + r_i \boldsymbol{X}_N + \varepsilon\}.$$

By weak continuity of $\succsim_i$, O^ε is open. Since $(d^k, r^k, r'^k)_{k \in \mathbb{N}}$ converges to (d,r,r'), it follows that $(d_i^k + r_i^k \boldsymbol{X}_N)_{k \in \mathbb{N}}$ weakly converges to $d_i + r_i \boldsymbol{X}$. Hence, there exists $K^\varepsilon \in \mathbb{N}$ such that $d_i^k + r_i^k \boldsymbol{X}_N \in O^\varepsilon$ for all $k \geq K^\varepsilon$. This implies that $f_i(d^k, r^k, r'^k) + r'^k \boldsymbol{X}_S \in O^\varepsilon$ for all $k \geq K^\varepsilon$. Since $\varepsilon > 0$ is arbitrary, it follows that $\lim_{k \to \infty} \left(f_i(d^k, r^k, r'^k) + r'^k \boldsymbol{X}_S\right) \sim_i d_i + r_i \boldsymbol{X}_N$. Since $\lim_{k \to \infty} r'^k = r'$, we obtain that $\lim_{k \to \infty} \left(f_i(d^k, r^k, r'^k)\right) + r' \boldsymbol{X}_S \sim_i d_i + r_i \boldsymbol{X}_N$. Condition (C2) and $f_i(d,r,r') + r'_i \boldsymbol{X}_S \sim_i d_i + r_i \boldsymbol{X}_N$ then imply that $\lim_{k \to \infty} f_i(d^k, r^k, r'^k) = f_i(d,r,r')$. Hence, f_i is continuous in (d,r,r'). □

5.2 Nucleolus $\mathcal{N}^2$

In the previous section, the excess function is based on adjusting a coalition's random payoff by adding or subtracting a deterministic amount of money. We obtain another excess function if we adjust the allocation $(d,r) \in IR_\Gamma(N)$ instead of the random payoffs $\boldsymbol{X}_S$. To see how this (adjustment) procedure works, suppose that coalition S has no incentive to part company with the grand coalition if the allocation $(d,r) \in IR_\Gamma(N)$ is proposed, that is there exists no allocation $(d',r') \in PO_\Gamma(S)$ such that $d'_i + r'_i\boldsymbol{X}_S \succ_i d_i + r_i\boldsymbol{X}_N$ for all $i \in S$. Then (d,r) allocates more to coalition S than they can obtain on their own. Hence, we can allocate even less to this coalition without providing them with an incentive to leave the grand coalition. More specifically, we can take away the amount $\sum_{i\in S}\delta_i < 0$ from coalition S, if there exists a Pareto optimal allocation $(d',r') \in PO_\Gamma(S)$ such that $d'_i + r'_iX_S \sim_i d_i + \delta_i + r_i\boldsymbol{X}_N$ for all $i \in S$. Moreover, the maximum amount of money we can take away in this way from coalition S, equals

$$\min\left\{\sum_{i\in S}\delta_i \,\middle|\, \exists_{(d',r')\in PO_\Gamma(S)}\forall_{i\in S} : d'_i + r'_i\boldsymbol{X}_S \sim_i d_i + \delta_i + r_i\boldsymbol{X}_N\right\}. \tag{5.4}$$

Defining the excess function as such, however, is unsuitable for the following reason.

Consider Figure 5.3 that depicts the expected utility levels that coalition $S = \{i,j\}$ can obtain in some stochastic cooperative game Γ, and let $(d^1,r^1) \in IR_\Gamma(N)$ be the allocation that is proposed by the grand coalition. Note that since the allocation (d^1,r^1) falls outside $IR_S(\Gamma)$, coalition S has no incentive to leave the grand coalition. If we allocate an amount of money x less to coalition S, the allocations they can obtain equal

$$\begin{aligned} B(x,(d^1,r^1)) = \; & \{(d^1+\delta,r^1) \in IS_\Gamma(N) | \\ & \exists_{\delta\in\mathbb{R}^N} : \textstyle\sum_{i\in S}\delta_i = x \text{ and } \forall_{i\notin S} : \delta_i = 0\}. \end{aligned} \tag{5.5}$$

Note that $(d^1,r^1) \in B(0,(d^1,r^1))$ and that $B(0,(d,r)) = B(0,(d^1,r^1))$ for all $(d,r) \in B(0,(d^1,r^1))$. Consequently, the maximum amount of money x^* that we can take away from coalition S when (d^1,r^1) is proposed by coalition N, is the same for all allocations $(d,r) \in B(0,(d^1,r^1))$.

Next, consider Figure 5.4. Again, coalition S has no incentive to part company with the grand coalition if the proposed allocation equals $(d^1,r^1) \in IR_\Gamma(N)$. In this case, however, an inconsistency arises. Recall that for any allocation $(d,r) \in B(0,(d^1,r^1))$ the maximum amount that can be taken away equals x^*. In particular, this holds for the allocation $(d^2,r^2) \in B(0,(d^1,r^1))$. But if the grand coalition would have agreed on the allocation (d^2,r^2), coalition S might threaten

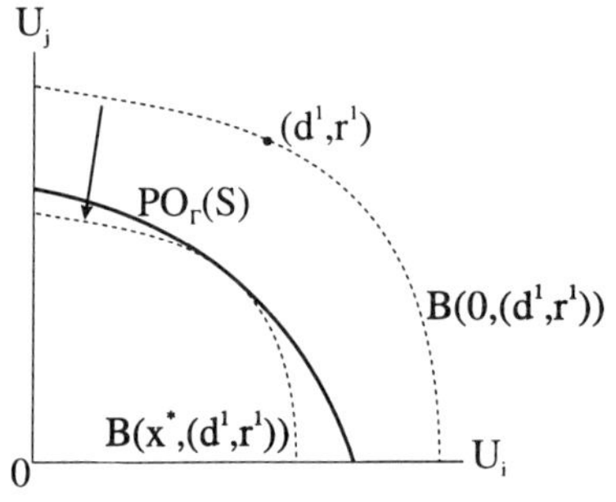

Figure 5.3

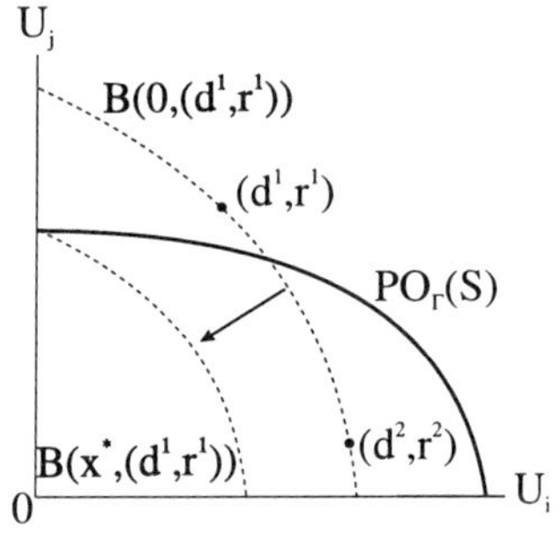

Figure 5.4

to leave the grand coalition, because they can obtain better payoffs on their own. Hence, coalition S would receive too little in the allocation (d^2, r^2), and one would expect that additional money is needed to make them stay in the grand coalition. The excess, however, states the opposite, namely that money can be taken away from them.

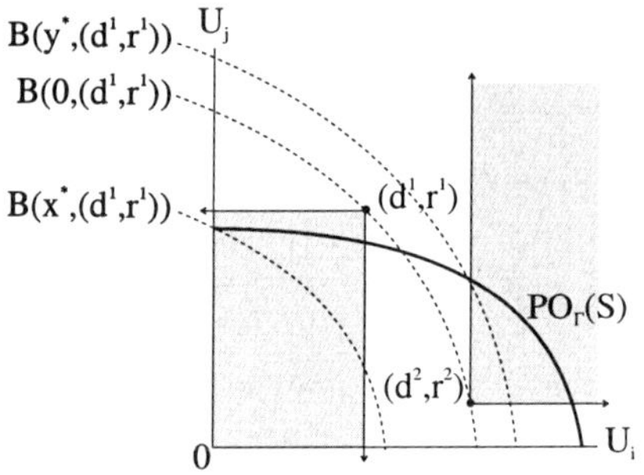

Figure 5.5

Summarizing, when defining the excess by expression (5.4), the counterintuitive and undesirable phenomenon may occur, in which a coalition exhibits the same complaints for two possible allocations, one that they can improve upon by leaving the grand coalition and cooperating on its own, and one that they cannot improve upon. To avoid this from happening, we adjust the definition of the excess in the following way. In case we would like to take away a certain amount of money from a coalition, we do so by taking away an amount of money from each individual member of this coalition. Similarly, if we would like to give a certain amount of money to a coalition, we do so by giving money to each individual member. As Figure 5.5 shows, this inconsistency is now resolved. For the excess of the allocation (d^1, r^1) is now determined by considering only those allocations $(d^1 + \delta, r^1)$ for which $\delta \leq 0$. Similarly, the excess of (d^2, r^2) is determined by considering only those allocations $(d^2 + \delta, r^2)$ for which $\delta \geq 0$. Consequently, the excess of the allocations (d^1, r^1) and (d^2, r^2) then equals $x^* < 0$ and $y^* > 0$, respectively.

5.2.1 Preliminary Definitions

In this section we go through some preliminaries that are necessary for the definition of the excess function E_S^2. Let $\Gamma \in CG(N)$ be a stochastic cooperative game. Define for each $S \subset N$

$$PD_\Gamma(S) = \{(d,r) \in I_\Gamma(S) | \; \exists_{(d',r') \in PO_\Gamma(S)} \forall_{i \in S} : d_i' + r_i' \boldsymbol{X}_S \succsim_i d_i + r_i \boldsymbol{X}_S\}$$

as the set of (possibly nonfeasible) allocations that are (weakly) dominated by a Pareto optimal allocation, and

$$NPD_\Gamma(S) = \{(d,r) \in I_\Gamma(S) | \; \nexists_{(d',r') \in PO_\Gamma(S)} \forall_{i \in S} : d_i' + r_i' \boldsymbol{X}_S \succsim_i d_i + r_i \boldsymbol{X}_S\}$$

as the set of (possibly nonfeasible) allocations that are not dominated by Pareto optimal allocations. Note that $IR_\Gamma(S) \subset PD_\Gamma(S)$. The reverse, however, need not be true, as the next example shows.

Example 5.4 Consider the following two person example. Let $\boldsymbol{X}_S$ be such that $-\boldsymbol{X}_S$ is exponentially distributed with expectation equal to 1 for all $S \subset N$. Furthermore let agents 1 and 2 be expected utility maximizers with utility functions $U_1(t) = -e^{-0.5t}$ and $U_2(t) = -e^{-0.25t}$, respectively. Then $E(U_1(d_1 + r_1 \boldsymbol{X}_{\{1,2\}})) = -e^{-d_1} \frac{1}{1-0.5r_1}$ and $E(U_2(d_2 - r_2 \boldsymbol{X}_{\{1,2\}})) = -e^{-d_2} \frac{1}{1-0.25r_2}$. An allocation $(d,r) \in I_\Gamma(\{1,2\})$ is individually rational if $E(U_1(d_1 + r_1 \boldsymbol{X}_{\{1,2\}})) \geq -2$ and $E(U_2(d_2 + r_2 \boldsymbol{X}_{\{1,2\}})) \geq -1.25$. Furthermore, (d, r^*) is Pareto optimal if and only if $r_1^* = \frac{1}{3}$ and $r_2^* = \frac{2}{3}$ (see Wilson (1968) or Proposition 6.2 for a similar result). Now, consider the allocation (d,r) with $d_1 = 0.1$, $d_2 = 0.1$, $r_1 = 1$ and $r_2 = 0$. Since $d_1 + d_2 > 0$ this allocation is nonfeasible. However, the Pareto optimal allocation (d^*, r^*) with $d_1^* = -0.9$, $d_2^* = 0.9$, $r_1^* = \frac{1}{3}$ and $r_2^* = \frac{2}{3}$ is feasible and preferred by both agents. Indeed,

$$E(U_1(d_1^* + r_1^* \boldsymbol{X}_{\{1,2\}})) = -1.8820 > -1.9025 = E(U_1(d_1 + r_1 \boldsymbol{X}_{\{1,2\}}))$$

and

$$E(U_2(d_2^* + r_2^* \boldsymbol{X}_{\{1,2\}})) = -0.9582 > -0.9753 = E(U_2(d_2 + r_2 \boldsymbol{X}_{\{1,2\}})).$$

So even nonfeasible allocations can be Pareto dominated.

The next proposition states a rather intuitive result. Namely that for every Pareto dominated allocation (d, r) and every non-Pareto dominated allocation (d', r'), which all members of S weakly prefer to the Pareto dominated allocation (d, r), there exists a Pareto optimal allocation such that for each agent the Pareto optimal allocation is weakly better than (d, r) but weakly worse than (d', r').

Proposition 5.5 Let $\Gamma \in CG(N)$. Take $(d, r) \in PD_\Gamma(S)$ and $(\tilde{d}, \tilde{r}) \in NPD_\Gamma(S)$ such that $d_i + r_i \boldsymbol{X}_S \precsim_i \tilde{d}_i + \tilde{r}_i \boldsymbol{X}_S$ for all $i \in S$. Then there exists $(\hat{d}, \hat{r}) \in PO_\Gamma(S)$ such that

$$d_i + r_i \boldsymbol{X}_S \precsim_i \hat{d}_i + \hat{r}_i \boldsymbol{X}_S \precsim_i \tilde{d}_i + \tilde{r}_i \boldsymbol{X}_S$$

for all $i \in S$.

PROOF: See Appendix 5.4. □

A direct consequence of this proposition is that for each allocation $(d, r) \in IR_\Gamma(S)$ there exists a Pareto optimal allocation (d', r') such that $d'_i + r'_i \boldsymbol{X}_S \succsim_i d_i + r_i \boldsymbol{X}_S$ for all $i \in S$. Moreover, since $IR_\Gamma(S)$ is assumed to be nonempty we have that for each $(d, r) \in NPD_\Gamma(S)$ there exists $(d', r') \in PO_\Gamma(S)$ such that $d'_i + r'_i \boldsymbol{X}_S \precsim_i d_i + r_i \boldsymbol{X}_S$ for all $i \in S$.

Finally, we introduce three more sets. Therefore, let $(d, r) \in IR_\Gamma(N)$ be an individually rational allocation for the grand coalition N. Take $S \subset N$ and define

$$W_S((d, r)) = \{(d', r') \in IR_\Gamma(S) | \forall_{i \in S} : d'_i + r'_i \boldsymbol{X}_S \precsim_i d_i + r_i \boldsymbol{X}_N\}$$

as the set of individually rational allocations for coalition S which are weakly worse than the payoff $d_i + r_i \boldsymbol{X}_N$ for every member of S, and,

$$B_S((d, r)) = \{(d', r') \in IR_\Gamma(S) | \forall_{i \in S} : d'_i + r'_i \boldsymbol{X}_S \succsim_i d_i + r_i \boldsymbol{X}_N\}$$

as the set of individually rational allocations for coalition S which are weakly better than the payoff $d_i + r_i \boldsymbol{X}_N$ for every member of S. Furthermore, define

$$PO_S^*((d, r)) = (W_S((d, r)) \cup B_S((d, r))) \cap PO_\Gamma(S),$$

as the set of Pareto optimal allocations for coalition S which are either weakly worse than $d_i + r_i \boldsymbol{X}_N$ for all members of S or weakly better than $d_i + r_i \boldsymbol{X}_N$ for all members of S. These three sets are illustrated in Figure 5.6 and Figure 5.7. Note that $B_S((d, r))$ can be empty.

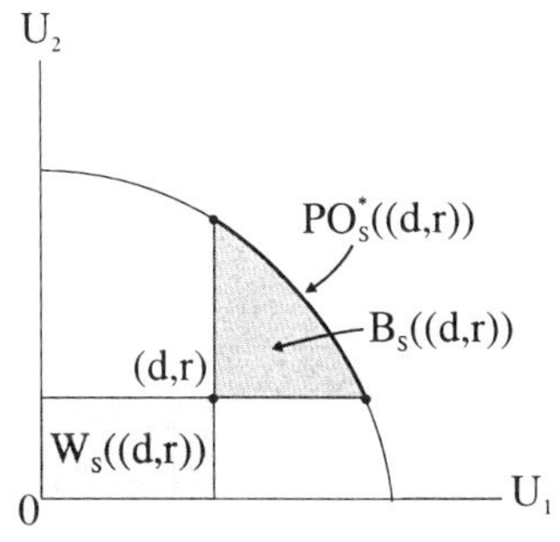

Figure 5.6

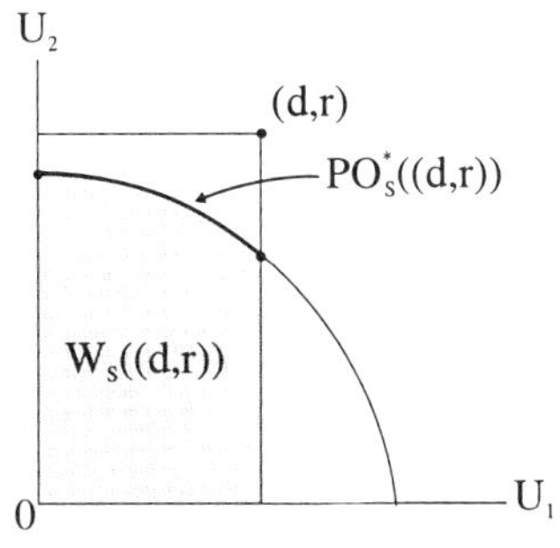

Figure 5.7

5.2.2 Definition of the Excess Function and Nucleolus

With the definitions and notions introduced in the previous section we can now adjust the excess function as defined in (5.4) in the appropriate way. Take $(d,r) \in IR_\Gamma(N)$. Then the excess function $E_S^2 : IR_\Gamma(N) \to \mathbb{R}$ for coalition S is defined by

$$E_S^2((d,r)) = \min_{(d',r') \in PO_S^*((d,r))} \{\sum_{i \in S} \delta_i | \; \forall_{i \in S} : \delta_i \in \mathbb{R} \text{ and } d_i' + r_i' \boldsymbol{X}_S \sim_i d_i + r_i \boldsymbol{X}_N + \delta_i\}.$$

Thus, the excess represents how much we can change the deterministic part d of an allocation $(d,r) \in IR_\Gamma(N)$, such that a coalition is indifferent between staying and leaving the grand coalition, while taking into account the additional requirement that the individual payoffs of all members change in the same direction.

Similar to the excess function E_S^1, if (d,r) and (d',r') are allocations of $\boldsymbol{X}_N$ such that each agent $i \in S$ prefers $d_i + r_i \boldsymbol{X}_N$ to $d_i' + r_i' \boldsymbol{X}_N$ then $E_S^2((d,r)) < E_S^2((d',r'))$. Hence, the excess decreases when each agent $i \in S$ improves his payoff. So, in a specific way the excess $E_S^2((d,r))$ describes how much coalition S is satisfied with the allocation (d,r). Since all agents' preferences are monotonically increasing in the amount of money d they receive (see assumption (C2)) it is reasonable to say that one coalition is more satisfied with a particular allocation than another coalition if the excess of the first coalition is less than the excess of the latter. Thus, the *nucleolus* $\mathcal{N}^2$ is defined as

$$\mathcal{N}^2(\Gamma) = \{(d,r) \in IR_\Gamma(N) | \; \forall_{(d',r') \in IR_\Gamma(N)} : \theta \circ E^2((d,r)) \leq_{lex} \theta \circ E^2((d',r'))\}, \tag{5.6}$$

where $E^2(d,r) = (E_S^2(d,r))_{S \subset N}$.

Theorem 5.6 Let Γ be a stochastic cooperative game. If $\Gamma \in CG(N)$ and $IR_\Gamma(N) \neq \emptyset$ then $\mathcal{N}(\Gamma) \neq \emptyset$.

PROOF: Similar to the proof of Theorem 5.3, we apply the results of Maschler et al. (1992). So, we have to show that $IR_\Gamma(N)$ is compact and that $E^2_S((d,r))$ is continuous in (d,r) for each $(d,r) \in IR_\Gamma(N)$ and each $S \subset N$. The compactness of $IR_\Gamma(N)$ follows immediately from Proposition 5.1. The continuity proof is a bit more complicated and consists of the following parts.

First, it follows from Lemma 5.12 in Appendix 5.4 that $PO^*_S((d,r))$ is a nonempty compact subset of $PO_\Gamma(S)$. Next, we introduce the multifunction

$$\overline{E}^2_S((d,r)) = \{\sum_{i\in S} \delta_i | \exists_{(d',r')\in PO^*_S((d,r))} : d'_i + r'_i \boldsymbol{X}_S \sim_i d_i + r_i \boldsymbol{X}_N + \delta_i\},$$

so that $E^2_S((d,r)) = \min \overline{E}^2_S((d,r))$. Lemma 5.13 shows that $\overline{E}^2_S((d,r))$ is a compact subset of $\mathbb{R}$ for each allocation $(d,r) \in IR_\Gamma(N)$, which implies that $\min \overline{E}^2_S((d,r))$ exists. Subsequently, Lemma 5.14 and Lemma 5.15 show that this multifunction is upper and lower semi continuous, respectively. From Lemma 5.16 it then follows that the excess function E^2_S is continuous. □

5.3 Nucleoli, Core, and Certainty Equivalents

For deterministic cooperative games it is known that the nucleolus as defined in Schmeidler (1969) is a core allocation whenever the core is nonempty. Similar results can be derived for the nucleoli $\mathcal{N}^k(\Gamma)$, $k = 1, 2$, introduced in this chapter. For this, recall that an allocation $(d,r) \in IR_\Gamma(N)$ belongs to the core $\mathcal{C}(\Gamma)$, if for each coalition $S \subset N$ there exists no allocation $(\bar{d},\bar{r}) \in IR_\Gamma(S)$ such that $\bar{d}_i + \bar{r}_i \boldsymbol{X}_S \succ_i d_i + r_i \boldsymbol{X}_N$ for all $i \in S$.

Theorem 5.7 Let $\Gamma \in CG(N)$. If $\mathcal{C}(\Gamma) \neq \emptyset$ then $\mathcal{N}^k(\Gamma) \subset \mathcal{C}(\Gamma)$ for $k = 1, 2$.

PROOF: We start with proving that $\mathcal{N}^1(\Gamma) \subset \mathcal{C}(\Gamma)$. Take $S \subset N$. First, we show that the excess $E^1_S(d,r) > 0$ if and only if there exists $(d',r') \in IR_\Gamma(S)$ that strongly dominates (d,r).

Let $(d,r) \in IR_\Gamma(N)$ be such that there exists $(d',r') \in IR_\Gamma(S)$ satisfying $d'_i + r'_i \boldsymbol{X}_S \succ_i d_i + r_i \boldsymbol{X}_N$ for all $i \in S$. By condition (C2), $d'_i - \delta_i + r'_i \boldsymbol{X}_S \sim_i d_i + r_i \boldsymbol{X}_N$ implies that $\delta_i > 0$. Hence, $E^1_S(d,r) \geq \sum_{i\in S} \delta_i > 0$.

For the reverse statement, let $(d',r') \in IR_\Gamma(S)$ be such that $\sum_{i\in S} \delta_i > 0$ if $d'_i - \delta_i + r'_i \boldsymbol{X}_S \sim_i d_i + r_i \boldsymbol{X}_N$ for all $i \in S$. Define $d^*_i = d'_i - \delta_i + \delta$ for all $i \in S$ with $\delta = \frac{1}{|S|} \sum_{i\in S} \delta_i > 0$. Since $\sum_{i\in S} d^*_i = 0$ it holds that $(d^*,r') \in IR_\Gamma(S)$.

Moreover, $d_i^* - \delta + r_i' \boldsymbol{X}_S \sim_i d_i + r_i \boldsymbol{X}_N$ for all $i \in S$, which implies that $d_i^* + r_i' \boldsymbol{X}_S \succ_i d_i + r_i \boldsymbol{X}_N$ for all $i \in S$. Hence, (d^*, r') strongly dominates (d, r).

Now, we show that the excess vector corresponding to a core allocation is lexicographically smaller than the excess vector corresponding to an allocation that does not belong to the core. This implies that the latter allocation cannot belong to the nucleolus of the game, whenever the core is nonempty. Hence, the nucleolus is a subset of the core.

Take $(d, r) \in \mathcal{C}(\Gamma)$ and $(d', r') \notin \mathcal{C}(\Gamma)$. Since $(d, r) \in \mathcal{C}(\Gamma)$ it follows that for each coalition $S \subset N$ there exists no allocation $(d^*, r^*) \in IR_\Gamma(S)$ that strictly dominates (d, r). Hence, $E_S^1(d, r) < 0$ for all $S \subset N$ with $S \neq N$. Since $(d', r') \notin \mathcal{C}(\Gamma)$ there exists a coalition $S \subset N$ and an allocation $(d^*, r^*) \in IR_\Gamma(S)$ such that (d^*, r^*) strictly dominates (d, r). Hence, $E_S^1(d, r) > 0$. This implies that $\theta \circ E^1(d, r) <_{lex} \theta \circ E^1(d', r')$, so that $(d', r') \notin \mathcal{N}^1(\Gamma)$.

In a similar way, we will now prove that $\mathcal{N}^2(\Gamma) \subset \mathcal{C}(\Gamma)$. For this, take $(d, r) \in IR_\Gamma(N)$ and $S \subset N$. Let $(\tilde{d}^S, \tilde{r}^S) \in I_\Gamma(S)$ be such that $\tilde{d}_i^S + \tilde{r}_i^S \boldsymbol{X}_S \sim_i d_i + r_i \boldsymbol{X}_N$ for all $i \in S$. Moreover, let $(\bar{d}, \bar{r}) \in PO_S^*((d, r))$ and $\delta \in \mathbb{R}^S$ be such that $\bar{d}_i + \bar{r}_i \boldsymbol{X}_S \sim_i d_i + r_i \boldsymbol{X}_N + \delta_i$ for all $i \in S$ and $\sum_{i \in S} \delta_i = E_S^2((d, r))$. Regarding the sign of the excess, we distinguish three cases.

First, suppose $(\tilde{d}^S, \tilde{r}^S) \in PD_\Gamma(S) \backslash PO_\Gamma(S)$. Then $\bar{d}_i + \bar{r}_i \boldsymbol{X}_S \succsim_i \tilde{d}_i^S + \tilde{r}_i^S \boldsymbol{X}_S$ for all $i \in S$. Hence, $\delta_i \geq 0$ for all $i \in S$. Since $(\tilde{d}^S, \tilde{r}^S)$ is not Pareto optimal there exists $j \in S$ such that $\bar{d}_j + \bar{r}_j \boldsymbol{X}_S \succ_j \tilde{d}_j^S + \tilde{r}_j^S \boldsymbol{X}_S \sim_j d_j + r_j \boldsymbol{X}_N$. Then $\delta_j > 0$ and, consequently, $E_S^2((d, r)) = \sum_{i \in S} \delta_i > 0$.

Second, suppose $(\tilde{d}^S, \tilde{r}^S) \in PO_\Gamma(S)$. This implies that $0 \in \overline{E}_S^2((d, r))$. Hence, $E_S^2((d, r)) \leq 0$.

Third, suppose $(\tilde{d}^S, \tilde{r}^S) \in NPD_\Gamma(S) \backslash PO_\Gamma(S)$. Then $\bar{d}_i + \bar{r}_i \boldsymbol{X}_S \precsim_i \tilde{d}_i^S + \tilde{r}_i^S \boldsymbol{X}_S$ for all $i \in S$. Hence, $\delta_i \leq 0$ for all $i \in S$. Moreover, since $(\tilde{d}^S, \tilde{r}^S)$ is not Pareto optimal there exists $j \in S$ such that $\bar{d}_j + \bar{r}_j \boldsymbol{X}_S \prec_j \tilde{d}_i^S + \tilde{r}_j^S \boldsymbol{X}_S \sim_j d_j + r_j \boldsymbol{X}_N$. So, $\delta_j < 0$ and, consequently, $E_S^2((d, r)) = \sum_{i \in S} \delta_i < 0$.

The excess vector corresponding to a core allocation is lexicographically smaller then the excess vector corresponding to an allocation that does not belong to the core. To see this, take $(d, r) \in \mathcal{C}(\Gamma)$ and $(d', r') \notin \mathcal{C}(\Gamma)$. Since $(d, r) \in \mathcal{C}(\Gamma)$ it follows from the core conditions that $(\tilde{d}^S, \tilde{r}^S) \in NPD_\Gamma(S)$ for all $S \subset N$. Hence, $E_S((d, r)) \leq 0$ for all $S \subset N$. Since $(d', r') \notin \mathcal{C}(\Gamma)$ there exists a coalition $S \subset N$ and an allocation $(\hat{d}, \hat{r}) \in IR_\Gamma(S)$ for S such that $\hat{d}_i + \hat{r}_i \boldsymbol{X}_S \succ_i d_i' + r_i' \boldsymbol{X}_N$ for all $i \in S$. Hence, $(\tilde{d}'^S, \tilde{r}'^S) \in PD_\Gamma(S) \backslash PO_\Gamma(S)$ and, consequently, $E_S((d', r')) > 0$. This implies that $\theta \circ E_S((d, r)) <_{lex} \theta \circ E_S((d', r'))$. Thus $(d', r') \notin \mathcal{N}(\Gamma)$. □

Consider the class $MG(N)$ of stochastic cooperative games introduced in Section 3.2.2. Recall that each stochastic cooperative game $\Gamma \in MG(N)$ gives rise to a TU-game (N, v_Γ). We will show that an allocation $(d,r) \in IR_\Gamma(N)$ belongs to the nucleolus $\mathcal{N}(\Gamma)$ if and only if the corresponding allocation $(m_i(d_i + r_i\boldsymbol{X}_N))_{i\in N}$ of certainty equivalents is the nucleolus of the corresponding TU-game (N, v_Γ).

Theorem 5.8 Let $\Gamma \in MG(N)$ such that $IR_\Gamma(N) \neq \emptyset$. Take $(d_i + r_i\boldsymbol{X}_N)_{i\in N} \in \mathcal{Z}_\Gamma(N)$ and let $y \in \mathbb{R}^N$ be such that $m_i(d_i + r_i\boldsymbol{X}_N) = y_i$ for all $i \in N$. Then $(d_i + r_i\boldsymbol{X}_N)_{i\in N} \in \mathcal{N}^k(\Gamma)$ for $k \in \{1,2\}$ if and only if $y = n(v_\Gamma)$.

PROOF: Let $\Gamma \in MG(N)$ and take $(d,r) \in IR_\Gamma(N)$. Let $y \in \mathbb{R}^N$ be such that $m_i(d_i + r_i\boldsymbol{X}_N) = y_i$ for all $i \in N$. The excess $E_S^k(d,r)$, $k = 1,2$ of coalition S then equals

$$
\begin{aligned}
&E_S^1(d,r)\\
&= \min_{(d',r')\in IR_\Gamma(S)} \{\sum_{i\in S} \delta_i | \forall_{i\in S} : d'_i - \delta_i + r'_i\boldsymbol{X}_S \sim_i d_i + r_i\boldsymbol{X}_N\}\\
&= \min_{(d',r')\in IR_\Gamma(S)} \{\sum_{i\in S} \delta_i | \forall_{i\in S} : m_i(d'_i - \delta_i + r'_i\boldsymbol{X}_S) = m_i(d_i + r_i\boldsymbol{X}_N)\}\\
&= \min_{(d',r')\in IR_\Gamma(S)} \{\sum_{i\in S} \delta_i | \forall_{i\in S} : m_i(d'_i + r'_i\boldsymbol{X}_S) - \delta_i = m_i(d_i + r_i\boldsymbol{X}_N)\}\\
&= \min_{(d',r')\in IR_\Gamma(S)} \sum_{i\in S} (m_i(d'_i + r'_i\boldsymbol{X}_S) - m_i(d_i + r_i\boldsymbol{X}_N))\\
&= v_\Gamma(S) - \sum_{i\in S} y_i,
\end{aligned}
$$

and

$$
\begin{aligned}
&E_S^2((d,r))\\
&= \min_{(d',r')\in PO_S^*((d,r))} \{\sum_{i\in S} \delta_i | \forall_{i\in S} : d'_i + r'_i\boldsymbol{X}_S \sim_i d_i + r_i\boldsymbol{X}_N + \delta_i\}\\
&= \min_{(d',r')\in PO_S^*((d,r))} \{\sum_{i\in S} \delta_i | \forall_{i\in S} : m_i(d'_i + r'_i\boldsymbol{X}_S) = m_i(d_i + r_i\boldsymbol{X}_N + \delta_i)\}\\
&= \min_{(d',r')\in PO_S^*((d,r))} \{\sum_{i\in S} \delta_i | \forall_{i\in S} : m_i(d'_i + r'_i\boldsymbol{X}_S) = \delta_i + m_i(d_i + r_i\boldsymbol{X}_N)\}\\
&= \min_{(d',r')\in PO_S^*((d,r))} \sum_{i\in S} (m_i(d'_i + r'_i\boldsymbol{X}_S) - m_i(d_i + r_i\boldsymbol{X}_N))\\
&= v_\Gamma(S) - \sum_{i\in S} y_i.
\end{aligned}
$$

So, both the excess $E_S^1(d,r)$ and $E_S^2(d,r)$ of coalition S equal the excess $E(S,y)$ introduced by Schmeidler (1969) of coalition S for the TU-game (N, v_Γ) (see page 27). Moreover, for each allocation $(d,r) \in PO_\Gamma(N)$ in Γ the vector $(m_i(d_i + r_i\boldsymbol{X}_N))_{i\in N}$ is an allocation of $v_\Gamma(N)$ in (N, v_Γ) and, vice versa, for each allocation x of $v_\Gamma(N)$ there exists an allocation $(d,r) \in PO_\Gamma(N)$ in Γ such that $m_i(d_i + r_i\boldsymbol{X}_N) = x_i$ for all $i \in N$. Hence, we obtain that an allocation $(d_i + r_i\boldsymbol{X}_N)_{i\in N}$ belongs to the nucleolus $\mathcal{N}^k(\Gamma)$ of the game Γ for $k \in \{1,2\}$ if and only if the corresponding allocation $(m_i(d_i + r_i\boldsymbol{X}_N))_{i\in N} = y$ equals the nucleolus $n(v_\Gamma)$ of the corresponding TU-game (N, v_Γ). □

A corollary of this result is that both nucleoli coincide on the class $MG(N)$ of stochastic cooperative games.

Example 5.9 Consider the following three person game Γ. Let $-\boldsymbol{X}_{\{i\}} \sim \text{Exp}(1)$ for $i = 1, 2, 3$ and let $\boldsymbol{X}_S = \sum_{i \in S} \boldsymbol{X}_{\{i\}}$ if $|S| \geq 2$. So, each agent individually faces a random cost which is exponentially distributed with expectation equal to 1. The cost of a coalition then equals the sum of the cost of the members of this coalition. Furthermore, all agents are expected utility maximizers with utility functions $U_1(t) = -e^{-0.5t}$, $U_2(t) = -e^{-0.33t}$ and $U_3(t) = -e^{-0.25t}$, respectively. For the certainty equivalent m_i it holds that $m_i(d_i + r_i\boldsymbol{X}_S) = U_i^{-1}(E(U_i(d_i + r_i\boldsymbol{X}_S)))$. For the corresponding TU-game (N, v_Γ) of Γ we then get $v_\Gamma(\{1\}) = -1.3863$, $v_\Gamma(\{2\}) = -1.2164$, $v_\Gamma(\{3\}) = -1.1507$, $v_\Gamma(\{1,2\}) = -2.2314$, $v_\Gamma(\{1,3\}) = -2.1878$, $v_\Gamma(\{2,3\}) = -2.1582$ and $v_\Gamma(\{1,2,3\}) = -3.1800$. The nucleolus $n(v_\Gamma)$ of this game is equal to $(-1.0933, -1.0633, -1.0234)$. To determine the nucleolus $\mathcal{N}(\Gamma)$ note that an allocation $(d_i + r_i\boldsymbol{X}_N)_{i \in N}\mathcal{Z}_\Gamma(N)$ is Pareto optimal if and only if $r = \frac{1}{9}(2, 3, 4)$. Then the only allocation $(d_i + r_i\boldsymbol{X}_N)_{i \in N}$ for which $(m_i(d_i + r_i\boldsymbol{X}_N))_{i \in N} = n(v_\Gamma)$ is the allocation $(d_i^* + r_i^*\boldsymbol{X}_N)_{i \in N}$ with $(d_1^*, d_2^*, d_3^*) = (-0.3865, -0.0034, 0.3899)$ and $(r_1^*, r_2^*, r_3^*) = \frac{1}{9}(2, 3, 4)$. Hence, $\mathcal{N}^1(\Gamma) = \mathcal{N}^2(\Gamma) = \{(d_i^* + r_i^*\boldsymbol{X}_N)_{i \in N}\}$.

5.4 Appendix: Proofs

Lemma 5.10 Let $\boldsymbol{X} \in L^1(\mathbb{R})$ and let $(d^k)_{k \in \mathbb{N}}$ and $(r^k)_{k \in \mathbb{N}}$ be sequences in $\mathbb{R}$ and $[0, 1]$, respectively. If $\lim_{k \to \infty} d^k = d$ and $\lim_{k \to \infty} r^k = r$ then $F_{d^k + r^k\boldsymbol{X}} \xrightarrow{w} F_{d + r\boldsymbol{X}}$.[2]

PROOF: We have to show that $\lim_{k \to \infty} F_{d^k + r^k\boldsymbol{X}}(t) = F_{d + r\boldsymbol{X}}(t)$ for all continuity points t of $F_{d + r\boldsymbol{X}}$. Note that $F_{d^k + r^k\boldsymbol{X}}(t) = F_{\boldsymbol{X}}(\frac{t - d^k}{r^k})$ if $r^k \neq 0$. We distinguish two cases, $r = 0$ and $r > 0$.

First, suppose that $r = 0$. Then $F_{d + r\boldsymbol{X}}(t) = 0$ if $t < d$ and $F_{d + r\boldsymbol{X}}(t) = 1$ if $t \geq d$. Take $t > d$ so that t is a continuity point of $F_{d + r\boldsymbol{X}}$. Let $\varepsilon > 0$. Since $d^k \to d$ there exists $K \in \mathbb{N}$ such that $t - d^k > \varepsilon$ for $k \geq K$. Since $r^k \to 0$ we have that $\frac{t - d^k}{r^k} \to \infty$. Hence, $\lim_{k \to \infty} F(\frac{t - d^k}{r^k}) = 1$. Similarly, one can show that $\lim_{k \to \infty} F(\frac{t - d^k}{r^k}) = 0$ if $t < d$.

Next, suppose that $r \neq 0$. Let $t \in \mathbb{R}$ be such that $\frac{t - d}{r}$ is a continuity point of $F_{\boldsymbol{X}}$. Since $\frac{t - d^k}{r^k} \to \frac{t - d}{r}$ it follows that $\lim_{k \to \infty} F(\frac{t - d^k}{r^k}) = F(\frac{t - d}{r})$. □

[2] See Appendix A for the definition of weak convergence '$\xrightarrow{w}$'.

Lemma 5.11 Let $\boldsymbol{X} \in L^1(\mathbb{R})$ and let $(d^k)_{k\in\mathbb{N}}$ and $(r^k)_{k\in\mathbb{N}}$ be sequences in $\mathbb{R}$ and $[0,1]$, respectively. If $F_{d^k+r^k\boldsymbol{X}} \xrightarrow{w} G$ then there exist numbers $d \in \mathbb{R}$ and $r \in [0,1]$ such that $F_{d^k+r^k\boldsymbol{X}} \xrightarrow{w} F_{d+r\boldsymbol{X}}$.

PROOF: Since $(r^k)_{k\in\mathbb{N}}$ is a sequence in the compact set $[0,1]$ there exists a convergent subsequence $(r^l)_{l\in\mathbb{N}}$. Let us denote its limit by $r \in [0,1]$. Lemma 5.10 then implies that $F_{r^l\boldsymbol{X}} \xrightarrow{w} F_{r\boldsymbol{X}}$. Since $F_{d^k+r^k\boldsymbol{X}} \xrightarrow{w} G$ we also have that $F_{d^l+r^l\boldsymbol{X}} \xrightarrow{w} G$. Next, we show that the sequence $(d^l)_{l\in\mathbb{N}}$ is bounded. This implies that the sequence $(d^l)_{l\in\mathbb{N}}$ has a convergent sequence $(d^m)_{m\in\mathbb{N}}$. If we denote its limit by $d \in \mathbb{R}$ it follows from Lemma 5.10 that $F_{d^m+r^m\boldsymbol{X}} \xrightarrow{w} F_{d+r\boldsymbol{X}}$. Since the limit G is unique, it also holds that $F_{d^k+r^k\boldsymbol{X}} \xrightarrow{w} F_{d+r\boldsymbol{X}}$.

So, we are left to prove that $(d^l)_{l\in\mathbb{N}}$ is a bounded sequence. Therefore, suppose that it is not bounded, then there exists a subsequence $(d_m)_{m\in\mathbb{N}}$ with either $d^m \to \infty$ or $d^m \to -\infty$.

Let us start with considering the first possibility, that is, $d^m \to \infty$. Note that without loss of generality we may assume that $d^1 \leq d^2 \leq d^3 \leq \ldots$. Take $t \in \mathbb{R}$ such that t is a continuity point of $F_{r\boldsymbol{X}}$ and $F_{r\boldsymbol{X}}(t) < 1$. Let $\varepsilon > 0$ be such that $F_{r\boldsymbol{X}}(t) + \varepsilon < 1$. Since $F_{r^m\boldsymbol{X}} \xrightarrow{w} F_{r\boldsymbol{X}}$ there exists $M_1 \in \mathbb{N}$ such that $F_{r^m\boldsymbol{X}}(t) < F_{r\boldsymbol{X}}(t) + \varepsilon$ for all $m \geq M_1$.

Let $\tau \in \mathbb{R}$ be a continuity point of G. Since $d^m \to \infty$ there exists $M_2 \geq M_1$ such that $t + d^m > \tau$ for all $m \geq M_2$. From $F_{r^m\boldsymbol{X}}(t) < F_{r\boldsymbol{X}}(t) + \varepsilon$ and $F_{d^m+r^m\boldsymbol{X}}(\tau) = F_{r^m\boldsymbol{X}}(\tau - d^m) < F_{r^m\boldsymbol{X}}(t)$ it then follows that $F_{d^m+r^m\boldsymbol{X}}(\tau) < F_{r\boldsymbol{X}}(t) + \varepsilon$ for all $m \geq M_2$. Hence, $\lim_{m\to\infty} F_{d^m+r^m\boldsymbol{X}}(\tau) = G(\tau)$ implies that $G(\tau) < F_{r\boldsymbol{X}}(t)+\varepsilon$. Since τ is an arbitrary continuity point of G and $F_{r\boldsymbol{X}}(t)+\varepsilon < 1$ it holds that $\lim_{\tau\to\infty} G(\tau) < 1$. But this contradicts the fact that G is a probability distribution function.

Next, consider the case that $d^m \to -\infty$. Again, we may assume without loss of generality that the sequencing is monotonic, i.e., $d^1 \geq d^2 \geq d^3 \geq \ldots$. Take $t \in \mathbb{R}$ such that t is a continuity point of $F_{r\boldsymbol{X}}$ and $F_{r\boldsymbol{X}}(t) > 0$. Let $\varepsilon > 0$ be such that $F_{r\boldsymbol{X}}(t) - \varepsilon > 0$. Since $F_{r^m\boldsymbol{X}} \xrightarrow{w} F_{r\boldsymbol{X}}$ there exists $M_1 \in \mathbb{N}$ such that $F_{r^m\boldsymbol{X}}(t) > F_{r\boldsymbol{X}}(t) - \varepsilon$ for all $m \geq M_1$.

Let $\tau \in \mathbb{R}$ be a continuity point of G. Since $d^m \to -\infty$ there exists $M_2 \geq M_1$ such that $t + d^m < \tau$ for all $m \geq M_2$. From $F_{r^m\boldsymbol{X}}(t) > F_{r\boldsymbol{X}}(t) - \varepsilon$ and $F_{d^m+r^m\boldsymbol{X}}(\tau) = F_{r^m\boldsymbol{X}}(\tau - d^m) > F_{r^m\boldsymbol{X}}(t)$ it then follows that $F_{d^m+r^m\boldsymbol{X}}(\tau) > F_{r\boldsymbol{X}}(t) - \varepsilon$ for all $m \geq M_2$. Hence, $\lim_{m\to\infty} F_{d^m+r^m\boldsymbol{X}}(\tau) = G(\tau)$ implies that $G(\tau) > F_{r\boldsymbol{X}}(t)-\varepsilon$. Since τ is an arbitrary continuity point of G and $F_{r\boldsymbol{X}}(t)-\varepsilon > 0$ it holds that $\lim_{\tau\to-\infty} G(\tau) > 0$, which contradicts the fact that G is a probability distribution function. □

Proposition 5.1 $IR_\Gamma(S)$ is a compact subset of $I_\Gamma(S)$ for each coalition $S \subset N$.

PROOF: Since $IR_\Gamma(S) \subset I_\Gamma(S) \subset \mathbb{R}^S \times \mathbb{R}^S$ it is sufficient to prove that $IR_\Gamma(S)$ is closed and bounded in $\mathbb{R}^S \times \mathbb{R}^S$. Since

$$IR_\Gamma(S) = \{(d,r) \in I_\Gamma(S) | \sum_{i \in S} d_i \leq 0\}$$

and $I_\Gamma(S)$ is closed by the weak continuity of $\succsim_i$ for all $i \in S$ it follows that $IR_\Gamma(S)$ is closed. To see that $IR_\Gamma(S)$ is bounded, define for each $i \in S$ and each $r_i \in [0,1]$

$$\underline{d}_i(r_i) = \min\{d_i | d_i + r_i \boldsymbol{X}_S \succsim_i \boldsymbol{X}_{\{i\}}\}.$$

Note that $\underline{d}_i(r_i)$ exists by assumptions (C5) and (C3) and that $\underline{d}_i(r_i) + r_i \boldsymbol{X}_S \sim_i \boldsymbol{X}_{\{i\}}$. To show that $\min_{r_i \in [0,1]} \underline{d}_i(r_i)$ exists it suffices to show that $\underline{d}_i(r_i)$ is continuous in r_i. Therefore, consider the sequence $(r_i^k)_{k \in \mathbb{N}}$ with $r_i^k \in [0,1]$ and $\lim_{k\to\infty} r_i^k = r_i$. By definition we have for all $k \in \mathbb{N}$ that $\underline{d}_i(r_i^k) + r_i^k \boldsymbol{X}_S \sim_i \boldsymbol{X}_{\{i\}}$. Hence, $\underline{d}_i(r_i^k) + r_i^k \boldsymbol{X}_S \sim_i \underline{d}_i(r_i) + r_i \boldsymbol{X}_S$ for all $i \in S$. Since $\succsim_i$ is weakly continuous it follows that

$$\lim_{k\to\infty} \left(\underline{d}_i(r_i^k) + r_i^k \boldsymbol{X}_S\right) = \lim_{k\to\infty} \underline{d}_i(r_i^k) + r_i \boldsymbol{X}_S \sim_i \underline{d}_i(r_i) + r_i \boldsymbol{X}_S$$

Then assumption (C2) implies that $\lim_{k\to\infty} \underline{d}_i(r_i^k) = \underline{d}_i(r_i)$. Consequently, $\underline{d}_i(r_i)$ is continuous in r_i and

$$\underline{d}_i = \min_{r_i \in [0,1]} \underline{d}_i(r_i)$$

exists and is finite for all $i \in S$.

Since $(d,r) \in IR_\Gamma(S)$ implies that $d_i + r_i \boldsymbol{X}_S \succsim_i \boldsymbol{X}_{\{i\}}$ for all $i \in S$ it follows by condition (C2) that $d_i \geq \underline{d}_i$ for all $i \in S$. Hence, $(d,r) \in IR_\Gamma(S)$ implies that

$$d \in \{\tilde{d} \in \mathbb{R}^S | \forall_{i \in S} : \tilde{d}_i \geq \underline{d}_i, \sum_{i \in S} \tilde{d}_i \leq 0\}$$

and $r \in \Delta^S$. Since both sets are bounded, we have that $IR_\Gamma(S)$ is bounded. □

Proposition 5.2 The set of Pareto optimal allocations $PO_\Gamma(S)$ is a compact subset of $I_\Gamma(S)$ for each coalition $S \subset N$.

PROOF: Since $PO_\Gamma(S) \subset IR_\Gamma(S)$ and $IR_\Gamma(S)$ is compact it is sufficient to show that $PO_\Gamma(S)$ is closed in $IR_\Gamma(S)$. Let $(d,r) \in IR_\Gamma(S)$ be such that $(d,r) \notin PO_\Gamma(S)$. Then there exists $(\bar{d},\bar{r}) \in IR_\Gamma(S)$ such that $\bar{d}_i + \bar{r}_i \boldsymbol{X}_S \succ_i d_i + r_i \boldsymbol{X}_S$

for all $i \in S$. Next, consider the set $\{(d',r') \in IR_\Gamma(S) | d'_i + r'_i \boldsymbol{X}_S \prec_i \bar{d}_i + \bar{r}_i \boldsymbol{X}_S\}$. By the weak continuity of $\succsim_i$ this set is open in $IR_\Gamma(S)$. Indeed, by the continuity of $\succsim_i$ we have that $\{\boldsymbol{Y} \in \prod_{i\in S} \mathcal{L}(\mathcal{X}_i) | \exists_{i\in S} : \boldsymbol{Y}_i \succsim_i \bar{d}_i + \bar{r}_i \boldsymbol{X}_S\}$ is closed. Hence, Proposition 5.1 implies that $\{(d',r') \in IR_\Gamma(S) | \exists_{i\in S} : d'_i + r'_i \boldsymbol{X}_S \succsim_i \bar{d}_i + \bar{r}_i \boldsymbol{X}_S\}$ is closed in $I_\Gamma(S)$. Hence, it is also closed in $IR_\Gamma(S)$. Consequently, $\{(d',r') \in IR_\Gamma(S) | \forall_{i\in S} : d'_i + r'_i \boldsymbol{X}_S \prec_i \bar{d}_i + \bar{r}_i \boldsymbol{X}_S\}$ must be open in $IR_\Gamma(S)$. Since (d,r) belongs to the latter set there exists an open neighborhood O of (d,r) in $IR_\Gamma(S)$ such that $O \subset \{(d',r') \in IR_\Gamma(S) | \forall_{i\in S} : d'_i + r'_i \boldsymbol{X}_S \prec_i \bar{d}_i + \bar{r}_i \boldsymbol{X}_S\}$. This implies that $(\tilde{d},\tilde{r}) \notin PO_\Gamma(S)$ whenever $(\tilde{d},\tilde{r}) \in O$. Hence, $IR_\Gamma(S) \backslash PO_\Gamma(S)$ is open in $IR_\Gamma(S)$ and, consequently, $PO_\Gamma(S)$ is closed in $IR_\Gamma(S)$. □

Proposition 5.5 Let $\Gamma \in CG(N)$. Take $(d,r) \in PD_\Gamma(S)$ and $(\tilde{d},\tilde{r}) \in NPD_\Gamma(S)$ such that $d_i + r_i \boldsymbol{X}_S \precsim_i \tilde{d}_i + \tilde{r}_i \boldsymbol{X}_S$ for all $i \in S$. Then there exists $(\hat{d},\hat{r}) \in PO_\Gamma(S)$ such that

$$d_i + r_i \boldsymbol{X}_S \precsim_i \hat{d}_i + \hat{r}_i \boldsymbol{X}_S \precsim_i \tilde{d}_i + \tilde{r}_i \boldsymbol{X}_S$$

for all $i \in S$.

PROOF: Let $(d,r) \in PD_\Gamma(S)$ and $(\tilde{d},\tilde{r}) \in NPD_\Gamma(S)$. Without loss of generality we may assume that $(d,r) \in IR_\Gamma(S)$.[3] Take $\delta_i \in \mathbb{R}$ be such that $d_i + \delta_i + r_i \boldsymbol{X}_S \sim_i \tilde{d}_i + \tilde{r}_i \boldsymbol{X}_S$. Note that $\delta_i \geq 0$ by condition (C2). Next, take $\bar{r} \in \Delta^S$ and $t \in [0,1]$. Let $\bar{d}_i(\bar{r},t)$ be such that $\bar{d}_i(\bar{r},t) + \bar{r}_i \boldsymbol{X}_S \sim_i d_i + t\delta_i + r_i \boldsymbol{X}_S$. Note that the allocation $(\bar{d}(\bar{r},t),\bar{r})$ is feasible if and only if $\sum_{i\in S} \bar{d}_i(\bar{r},t) \leq 0$. First, we show that $\bar{d}_i(\bar{r},t)$ is continuous in $(\bar{r},t)$. Let $((\bar{r}^k,t^k))_{k\in\mathbb{N}}$ be a convergent sequence with limit $(\bar{r},t)$. We have to show that $\lim_{k\to\infty} \bar{d}_i(\bar{r}^k,t^k) = \bar{d}_i(\bar{r},t)$. Note that $\bar{d}_i(\bar{r}^k,t^k) + \bar{r}^k_i \boldsymbol{X}_S \sim_i d_i + t^k\delta_i + r_i \boldsymbol{X}_S$ for all $k \in \mathbb{N}$. Define for $\varepsilon > 0$

$$\mathcal{O}^\varepsilon_i = \{\boldsymbol{Y} \in \mathcal{L}(\mathcal{X}_i) | d_i + t\delta_i + r_i \boldsymbol{X}_S - \varepsilon \prec_i \boldsymbol{Y} \prec_i d_i + t\delta_i + r_i \boldsymbol{X}_S + \varepsilon\}.$$

Since $t^k \to t$ there exists $K^\varepsilon \in \mathbb{N}$ such that $d_i + t^k\delta_i + r_i \boldsymbol{X}_S \in \mathcal{O}^\varepsilon_i$ for all $k > K^\varepsilon$. Consequently, we have that $\bar{d}_i(\bar{r}^k,t^k) + \bar{r}^k_i \boldsymbol{X}_S \in \mathcal{O}^\varepsilon_i$ for all $k > K^\varepsilon$. This implies that $\lim_{k\to\infty} \left(\bar{d}_i(\bar{r}^k,t^k) + \bar{r}^k_i \boldsymbol{X}_S\right) \in \cap_{\varepsilon>0} \mathcal{O}^\varepsilon_i$. So,

$$\lim_{k\to\infty} \left(\bar{d}_i(\bar{r}^k,t^k) + \bar{r}^k_i \boldsymbol{X}_S\right) = \lim_{k\to\infty} \bar{d}_i(\bar{r}^k,t^k) + \bar{r}_i \boldsymbol{X}_S \sim_i d_i + t\delta_i + r_i \boldsymbol{X}_S.$$

[3] If $(d,r) \notin IR_\Gamma(S)$ then there exists $(d',r') \in IR_\Gamma(S)$ such that $d'_i + r'_i \boldsymbol{X}_S \succ_i d_i + r_i \boldsymbol{X}_S$ for all $i \in S$. If $\delta'_i < 0$ is such that $d'_i + \delta'_i + r'_i \boldsymbol{X}_S \sim_i d_i + r_i \boldsymbol{X}_S$ for all $i \in S$ then $(d' + \delta', r')$ is still a feasible allocation. Thus, $(d' + \delta', r') \in IR_\Gamma(S)$. Continuing the proof with the allocation (d,r) replaced by $(d' + \delta', r')$ would yield the same result.

Since $\bar{d}_i(\bar{r}, t) + \bar{r}_i \boldsymbol{X}_S \sim_i d_i + t\delta_i + r_i \boldsymbol{X}_S$ it follows from condition (C2) that $\lim_{k\to\infty} \bar{d}_i(\bar{r}^k, t^k) = \bar{d}_i(\bar{r}, t)$.

Next, define $f(t) = \min_{\bar{r}\in\Delta^S} \sum_{i\in S} \bar{d}_i(\bar{r}, t)$ for all $t \in [0, 1]$. Then f is a continuous function. Moreover, since $(d, r) \in IR_\Gamma(S)$ and $\bar{d}_i(r, 0) = d_i$ for all $i \in S$ it follows from the feasibility of (d, r) that $f(0) \leq \sum_{i\in S} \bar{d}_i(r, 0) = \sum_{i\in S} d_i \leq 0$. Furthermore, since $d_i + \delta_i + r_i \boldsymbol{X}_S \sim_i \tilde{d}_i + \tilde{r}_i \boldsymbol{X}_S$ for all $i \in S$ and $(\tilde{d}, \tilde{r}) \in NPD_\Gamma(S)$ it follows that $(d + \delta, r) \in NPD_\Gamma(S)$. This implies that $f(1) \geq 0$. For, if $f(1) < 0$ then there exists $r^* \in \Delta^S$ such that $\sum_{i\in S} \bar{d}_i(r^*, 1) < 0$ and $\bar{d}_i(r^*, 1) + r_i^* \boldsymbol{X}_S \sim_i d_i + \delta_i + r_i \boldsymbol{X}_S$ for all $i \in S$. Consequently, the allocation yielding the payoffs

$$\bar{d}_i(r^*, 1) - \tfrac{1}{|S|} \sum_{i\in S} \bar{d}_i(r^*, 1) + r_i^* \boldsymbol{X}_S$$

for each $i \in S$ is feasible and preferred to $d_i + \delta_i + r_i \boldsymbol{X}_S$ by all agents $i \in S$. Clearly, this contradicts the fact that $(d + \delta, r) \in NPD_\Gamma(S)$. Thus, $f(0) \leq 0 \leq f(1)$. The continuity of f then implies that there exists $\hat{t}$ such that $f(\hat{t}) = 0$.

Let $\hat{r} \in \Delta^S$ be such that $\sum_{i\in S} \bar{d}_i(\hat{r}, \hat{t}) = 0$. Then the allocation $(\bar{d}(\hat{r}, \hat{t}), \hat{r})$ is Pareto optimal. To see this, first note that $\sum_{i\in S} \bar{d}_i(\bar{r}, \hat{t}) \geq 0$ for all $\bar{r} \in \Delta^S$. Second, note that the definition of $\bar{d}_i(\bar{r}, t)$ implies that

$$\bar{d}_i(\hat{r}, \hat{t}) + \hat{r}_i \boldsymbol{X}_S \sim_i \bar{d}_i(\bar{r}, \hat{t}) + \bar{r}_i \boldsymbol{X}_S \tag{5.7}$$

for all $i \in S$ and all $\bar{r} \in \Delta^S$. Next, take $\bar{r} \in \Delta^S$. If $\sum_{i\in S} \bar{d}_i(\bar{r}, \hat{t}) > 0$ then the allocation $(\bar{d}(\bar{r}, \hat{t}), \bar{r})$ is not feasible. From expression (5.7) it then follows that there exists no feasible allocation $(\bar{d}, \bar{r})$ which all agents $i \in S$ prefer to the allocation $(\bar{d}(\hat{r}, \hat{t}), \hat{r})$. If $\sum_{i\in S} \bar{d}_i(\bar{r}, \hat{t}) = 0$ then the allocation $(\bar{d}(\bar{r}, \hat{t}), \bar{r})$ is feasible. Moreover, an allocation $(\bar{d}, \bar{r})$ that all agents $i \in S$ prefer to $\bar{d}(\bar{r}, \hat{t}), \bar{r})$ must be infeasible by condition (C2) and expression (5.7). Hence, there exists no feasible allocation $(\bar{d}, \bar{r})$ which all agents $i \in S$ prefer to $(\bar{d}(\hat{r}, \hat{t}), \hat{r})$. Consequently, $(\bar{d}(\hat{r}, \hat{t}), \hat{r})$ is Pareto optimal. $0 \leq \hat{t} \leq 1$ then implies that

$$d_i + r_i \boldsymbol{X}_S \precsim_i \bar{d}_i(\hat{r}, \hat{t}) + \hat{r}_i \boldsymbol{X}_S \precsim_i d_i + \delta_i + r_i \boldsymbol{X}_S \sim_i \tilde{d}_i + \tilde{r}_i \boldsymbol{X}_S$$

for all $i \in S$. □

Lemma 5.12 $PO_S^*((d, r))$ is a nonempty compact subset of $PO_\Gamma(S)$.

PROOF: That $PO_S^*((d, r))$ is compact follows from the facts that $W_S((d, r))$ and $B_S((d, r))$ are closed by the continuity condition (C5) and $PO_\Gamma(S)$ is compact. To show that it is nonempty let us distinguish two cases.

First, let $B_S((d,r)) \neq \emptyset$. Then there exists $(d',r') \in IR_\Gamma(S)$ such that $d'_i + r'_i \boldsymbol{X}_S \succsim_i d_i + r_i \boldsymbol{X}_S$ for all $i \in S$. Since $(d',r') \in IR_\Gamma(S)$ we know from Proposition 5.5 that there exists $(\bar{d},\bar{r}) \in PO_\Gamma(S)$ such that $\bar{d}_i + \bar{r}_i \boldsymbol{X}_S \succsim_i d'_i + r'_i \boldsymbol{X}_S$ for all $i \in S$. Hence, $(\bar{d},\bar{r}) \in PO_\Gamma(S)$ and $(\bar{d},\bar{r}) \in B_S((d,r))$. Consequently, $(\bar{d},\bar{r}) \in PO^*_S((d,r))$.

Second, let $B_S((d,r)) = \emptyset$. Take $(\tilde{d},\tilde{r}) \in I_\Gamma(S)$ such that $\tilde{d}_i + \tilde{r}_i \boldsymbol{X}_S \sim_i d_i + r_i \boldsymbol{X}_N$ for all $i \in S$. From $B_S((d,r)) = \emptyset$ it follows that $(\tilde{d},\tilde{r}) \in NPD_\Gamma(S)$. Proposition 5.5 then implies that there exists $(\bar{d},\bar{r}) \in PO_\Gamma(S)$ such that $\bar{d}_i + \bar{r}_i \boldsymbol{X}_S \precsim_i \tilde{d}_i + \tilde{r}_i \boldsymbol{X}_S$ for all $i \in S$. Hence, $(\bar{d},\bar{r}) \in W_S((d,r))$ and, consequently, $(\bar{d},\bar{r}) \in PO^*_S((d,r))$. □

Lemma 5.13 Let $(d,r) \in IR_\Gamma(N)$. Then $\overline{E}^2_S((d,r))$ is a compact subset of $\mathbb{R}$.

PROOF: We have to show that $\overline{E}^2_S((d,r))$ is closed and bounded. That $\overline{E}^2_S((d,r))$ is bounded follows from the compactness of $PO^*_S((d,r))$ and the fact that for each $(d',r') \in PO^*_S((d,r))$ the number δ_i is uniquely determined by conditions (C3) and (C5). To see that $\overline{E}^2_S((d,r))$ is closed, let $(\sum_{i\in S} \delta_i^k)_{k\in\mathbb{N}}$ be a convergent sequence[4] in $\overline{E}^2_S((d,r))$ with limit $\sum_{i\in S}\delta_i$. We have to show that $\sum_{i\in S}\delta_i \in \overline{E}^2_S((d,r))$. Therefore, let $((\bar{d}^k,\bar{r}^k))_{k\in\mathbb{N}}$ be a sequence in $PO^*_S((d,r))$ such that $\bar{d}^k_i + \bar{r}^k_i \boldsymbol{X}_S \sim_i d_i + \delta_i^k + r_i \boldsymbol{X}_N$ for all $i \in S$. Since $PO^*_S((d,r))$ is compact there exists a convergent subsequence $((\bar{d}^l,\bar{r}^l))_{l\in\mathbb{N}}$ with limit $(\bar{d},\bar{r}) \in PO^*_S((d,r))$. Take $\bar{\delta}_i \in \mathbb{R}$ such that $\bar{d}_i + \bar{r}_i \boldsymbol{X}_S \sim_i d_i + \bar{\delta}_i + r_i \boldsymbol{X}_N$ for all $i \in S$. Note that $\sum_{i\in S}\bar{\delta}_i \in \overline{E}^2_S((d,r))$. The proof is finished if we can show that $\delta_i = \bar{\delta}_i$ for all $i \in S$. Therefore, let $\varepsilon > 0$ and $i \in S$. Define

$$\mathcal{O}_i^\varepsilon = \{\boldsymbol{Y} \in \mathcal{L}(\mathcal{X}_i) | \bar{d}_i + \bar{r}_i \boldsymbol{X}_S - \varepsilon \prec_i \boldsymbol{Y} \prec_i \bar{d}_i + \bar{r}_i \boldsymbol{X}_S + \varepsilon\}.$$

Since $\mathcal{O}_i^\varepsilon$ is open by the weak continuity of $\succsim_i$, $(\bar{d}^l,\bar{r}^l) \to (\bar{d},\bar{r})$ and $\bar{d}_i + \bar{r}_i \boldsymbol{X}_S \in \mathcal{O}_i^\varepsilon$ there exists $L^\varepsilon \in \mathbb{N}$ such that $\bar{d}^l_i + \bar{r}^l_i \boldsymbol{X}_S \in \mathcal{O}_i^\varepsilon$ for all $l > L^\varepsilon$. This implies that $d_i + \delta_i^l + r_i \boldsymbol{X}_N \in \mathcal{O}_i^\varepsilon$ for all $l > L^\varepsilon$. Since $\varepsilon > 0$ was arbitrarily chosen it follows that

$$\lim_{l\to\infty} \left(d_i + \delta_i^l + r_i^l \boldsymbol{X}_N\right) = d_i + \delta_i + r_i \boldsymbol{X}_N \in \cap_{\varepsilon>0} \mathcal{O}_i^\varepsilon.$$

Hence, $d_i + \delta_i + r_i \boldsymbol{X}_N \sim_i \bar{d}_i + \bar{r}_i \boldsymbol{X}_S$. Since by definition it holds that $\bar{d}_i + \bar{r}_i \boldsymbol{X}_S \sim_i d_i + \bar{\delta}_i + r_i \boldsymbol{X}_N$ it follows by assumption (C3) that $\delta_i = \bar{\delta}_i$. □

[4]Formally, it would be more correct to start with a convergent sequence $(a^k)_{k\in N}$ in $\overline{E}^2_S((d,r))$. Then $a^k \in \overline{E}^2_S((d,r))$ and the definition of $\overline{E}^2_S$ imply that there exist δ_i^k such that $d_i + \delta_i^k + r_i \boldsymbol{X}_N \sim_i d'_i + r'_i \boldsymbol{X}_S$ $(i \in S)$ for some $(d',r') \in PO^*_S((d,r))$ and $\sum_{i\in S}\delta_i^k = a^k$. Consequently, the sequence $(a^k)_{k\in\mathbb{N}}$ can be replaced by a sequence $(\sum_{i\in S}\delta_i^k)_{k\in\mathbb{N}}$.

Lemma 5.14 $\overline{E}^2_S((d,r))$ is upper semi continuous in (d,r) for all $(d,r) \in IR_\Gamma(N)$.

PROOF: Let $((d^k,r^k))_{k\in\mathbb{N}}$ be a sequence in $IR_\Gamma(N)$ converging to (d,r). Take $\sum_{i\in S}\delta_i^k \in \overline{E}^2_S((d^k,r^k))$ such that $\sum_{i\in S}\delta_i^k$ converges to $\sum_{i\in S}\delta_i$. For upper semi continuity to be satisfied it is sufficient to show that $\sum_{i\in S}\delta_i \in \overline{E}^2_S((d,r))$.

First, take $(\bar{d}^k,\bar{r}^k) \in PO^*_S((d^k,r^k))$ such that $\bar{d}_i^k + \bar{r}_i^k \boldsymbol{X}_S \sim_i d_i^k + \delta_i^k + r_i^k \boldsymbol{X}_N$ for all $i \in S$. Since $((\bar{d}^k,\bar{r}^k))_{k\in\mathbb{N}}$ is a sequence in the compact set $PO_\Gamma(S)$ there exists a convergent subsequence $((\bar{d}^l,\bar{r}^l))_{l\in\mathbb{N}}$ with limit $(\bar{d},\bar{r}) \in PO_\Gamma(S)$. Moreover, it holds that $\bar{d}_i + \bar{r}_i\boldsymbol{X}_S \sim_i d_i + \delta_i + r_i\boldsymbol{X}_N$ for all $i \in S$. To see this, take $\varepsilon > 0$ and $i \in S$. Define

$$\mathcal{O}_i^\varepsilon = \{\boldsymbol{Y} \in \mathcal{L}(\mathcal{X}_i) | \bar{d}_i + \bar{r}_i\boldsymbol{X}_S - \varepsilon \prec_i Y \prec_i \bar{d}_i + \bar{r}_i\boldsymbol{X}_S + \varepsilon\}.$$

Since $\mathcal{O}_i^\varepsilon$ is open by the continuity of $\succsim_i$, $(\bar{d}^l,\bar{r}^l) \to (\bar{d},\bar{r})$ and $\bar{d}_i + \bar{r}_i\boldsymbol{X}_S \in \mathcal{O}_i^\varepsilon$ there exists $L^\varepsilon \in \mathbb{N}$ such that $\bar{d}_i^l + \bar{r}_i^l\boldsymbol{X}_S \in \mathcal{O}_i^\varepsilon$ for all $l > L^\varepsilon$. This implies that $d_i^l + \delta_i^l + r_i^l\boldsymbol{X}_N \in \mathcal{O}_i^\varepsilon$ for all $l > L^\varepsilon$. Since $\varepsilon > 0$ is arbitrary we have that

$$\lim_{l\to\infty}\left(d_i^l + \delta_i^l + r_i^l\boldsymbol{X}_N\right) = d_i + \delta_i + r_i\boldsymbol{X}_N \sim_i \bar{d}_i + \bar{r}_i\boldsymbol{X}_S.$$

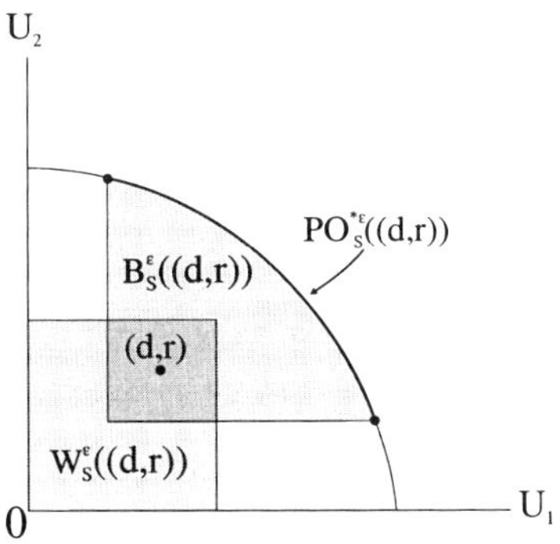

Figure 5.8

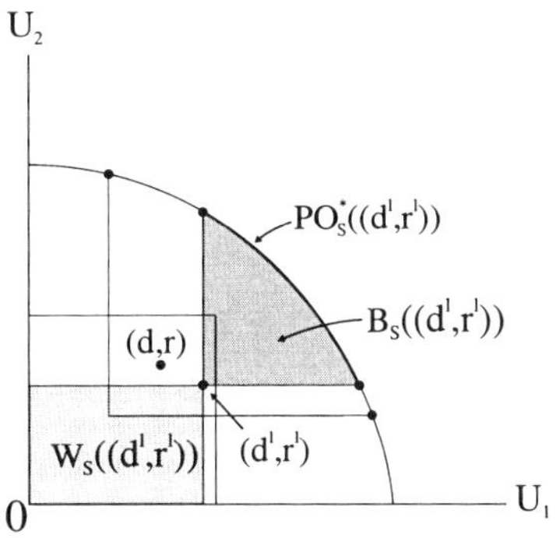

Figure 5.9

The proof is finished if we can show that $(\bar{d},\bar{r}) \in PO^*_S((d,r))$. Therefore, take $\varepsilon > 0$ and define

$$\begin{aligned} W_S^\varepsilon((d,r)) &= \{(d',r') \in IR_\Gamma(S) | \forall_{i\in S} : d_i' + r_i'\boldsymbol{X}_S \precsim_i d_i + r_i\boldsymbol{X}_N + \varepsilon\}, \\ B_S^\varepsilon((d,r)) &= \{(d',r') \in IR_\Gamma(S) | \forall_{i\in S} : d_i' + r_i'\boldsymbol{X}_S \succsim_i d_i + r_i\boldsymbol{X}_N - \varepsilon\}, \\ PO_S^{*\varepsilon}((d,r)) &= (W_S^\varepsilon((d,r)) \cup B_S^\varepsilon((d,r))) \cap PO_\Gamma(S). \end{aligned}$$

We refer to Figure 5.8 for a graphical interpretation of these sets. Note that

$W_S((d,r)) = \cap_{\varepsilon>0} W_S^{\varepsilon}((d,r))$, $B_S((d,r)) = \cap_{\varepsilon>0} B_S^{\varepsilon}((d,r))$ and $PO_S^*((d,r)) = \cap_{\varepsilon>0} PO_S^{*\varepsilon}((d,r))$. Furthermore, define

$$\mathcal{O}^{\varepsilon} = \{\boldsymbol{Y} \in \prod_{i\in S} \mathcal{L}(\mathcal{X}_i) | \forall_{i\in S} : d_i + r_i \boldsymbol{X}_N - \varepsilon \prec_i \boldsymbol{Y}_i \prec_i d_i + r_i \boldsymbol{X}_N + \varepsilon\}.$$

Since $\mathcal{O}^{\varepsilon}$ is open by the weak continuity of $\succsim_i$, $(d^l, r^l) \to (d,r)$ and $(d_i + r_i \boldsymbol{X}_N)_{i\in S} \in \mathcal{O}^{\varepsilon}$ there exists $L^{\varepsilon} \in \mathbb{N}$ such that $(d_i^l + r_i^l \boldsymbol{X}_N)_{i\in S} \in \mathcal{O}^{\varepsilon}$ for all $l > L^{\varepsilon}$. This implies that $(d^l, r^l) \in W_S^{\varepsilon}((d,r))$ and $(d^l, r^l) \in B_S^{\varepsilon}((d,r))$ for all $l > L^{\varepsilon}$. Hence, $W_S((d^l, r^l)) \subset W_S^{\varepsilon}((d,r))$ and $B_S((d^l, r^l)) \subset B_S^{\varepsilon}((d,r))$ for all $l > L^{\varepsilon}$ (see also Figure 5.9). Consequently, we have for all $l > L^{\varepsilon}$ that $PO_S^*((d^l, r^l)) \subset PO_S^{*\varepsilon}((d,r))$. In particular, we have $(\bar{d}^l, \bar{r}^l) \in PO_S^{*\varepsilon}((d,r))$ for all $l > L^{\varepsilon}$. Hence,

$$\lim_{l\to\infty} (\bar{d}^l, \bar{r}^l) = (\bar{d}, \bar{r}) \in \cap_{\varepsilon>0} PO_S^{*\varepsilon}((d,r)) = PO_S^*((d,r)). \qquad \square$$

Lemma 5.15 $\overline{E}_S^2((d,r))$ is lower semi continuous in (d,r) for all $(d,r) \in IR_{\Gamma}(N)$.

PROOF: Let $((d^k, r^k))_{k\in\mathbb{N}}$ be a sequence converging to (d,r) and let $\sum_{i\in S} \delta_i \in \overline{E}_S^2((d,r))$. To prove lower semi continuity it suffices to show that there exists a sequence $(\sum_{i\in S} \delta_i^k)_{k\in\mathbb{N}}$ with $\sum_{i\in S} \delta_i^k \in \overline{E}_S^2((d^k, r^k))$ for all $k \in \mathbb{N}$ such that $\sum_{i\in S} \delta_i^k$ converges to $\sum_{i\in S} \delta_i$.

First, note that since $IR_{\Gamma}(N)$ is compact and $\overline{E}_S^2$ is upper semi continuous that

$$\overline{E}_S^2(IR_{\Gamma}(N)) = \cup_{(d,r)\in IR_{\Gamma}(N)} E_S((d,r))$$

is a compact subset of $\mathbb{R}$. Second, note that if there exists a sequence $(\sum_{i\in S} \delta_i^k)_{k\in\mathbb{N}}$ with $\sum_{i\in S} \delta_i^k \in \overline{E}_S^2((d^k, r^k))$ for each $k \in \mathbb{N}$ such that every convergent subsequence $(\sum_{i\in S} \delta_i^l)_{l\in\mathbb{N}}$ converges to $\sum_{i\in S} \delta_i$ then the compactness of $\overline{E}_S^2(IR_{\Gamma}(N))$ implies that the sequence $(\sum_{i\in S} \delta_i^k)_{k\in\mathbb{N}}$ converges to $\sum_{i\in S} \delta_i$.

Take $(\bar{d}, \bar{r}) \in PO_S^*((d,r))$ such that $\bar{d}_i + \bar{r}_i \boldsymbol{X}_S \sim_i d_i + \delta_i + r_i \boldsymbol{X}_N$ for all $i \in S$. Let $\varepsilon > 0$ and define

$$\mathcal{O}^{\varepsilon} = \left\{ \boldsymbol{Y} \in \prod_{i\in S} \mathcal{L}(\mathcal{X}_i) \;\middle|\; \begin{array}{l} \forall_{i\in S} : \boldsymbol{Y}_i \succ_i \bar{d}_i + \bar{r}_i \boldsymbol{X}_S - \varepsilon \text{ or} \\ \forall_{i\in S} : \boldsymbol{Y}_i \prec_i \bar{d}_i + \bar{r}_i \boldsymbol{X}_S + \varepsilon \end{array} \right\}.$$

Note that $\mathcal{O}^{\varepsilon}$ is open by the weak continuity of $\succsim_i$ for each $i \in S$. Next, we show that if $(d_i^k + r_i^k \boldsymbol{X}_N)_{i\in S} \in \mathcal{O}^{\varepsilon}$ then there exists $(\bar{d}^k, \bar{r}^k) \in PO_S^*((d^k, r^k))$ such that $(\bar{d}_i^k + \bar{r}_i^k \boldsymbol{X}_S)_{i\in S} \in \mathcal{O}^{\varepsilon}$. Therefore, let $k \in \mathbb{N}$ be such that $(d^k, r^k) \in \mathcal{O}^{\varepsilon}$ and let $(\tilde{d}, \tilde{r}) \in I_{\Gamma}(S)$ be such that $\tilde{d}_i + \tilde{r}_i \boldsymbol{X}_S \sim_i d_i^k + r_i^k \boldsymbol{X}_N$ for all $i \in S$. We distinguish the following three cases.

First, suppose that $(\tilde{d},\tilde{r}) \in NPD_\Gamma(S)$. Since $(\tilde{d}_i+\tilde{r}_i\boldsymbol{X}_S)_{i\in S} \in \mathcal{O}^\varepsilon$ it holds that $\tilde{d}_i+\tilde{r}_i\boldsymbol{X}_S \succ_i \bar{d}_i+\bar{r}_i\boldsymbol{X}_S-\varepsilon$ for all $i \in S$. From $(\bar{d}-\frac{1}{2}(\varepsilon,\varepsilon,\ldots,\varepsilon),\bar{r}) \in PD_\Gamma(S)$ and Proposition 5.5 it follows that there exists $(\bar{d}^k,\bar{r}^k) \in PO_\Gamma(S)$ such that

$$\bar{d}_i+\bar{r}_i\boldsymbol{X}_S-\tfrac{1}{2}\varepsilon \precsim_i \bar{d}_i^k+\bar{r}_i^k\boldsymbol{X}_S \precsim_i \tilde{d}_i+\tilde{r}_i\boldsymbol{X}_S$$

for all $i \in S$. Thus, $(\bar{d}_i^k+\bar{r}_i^k\boldsymbol{X}_S)_{i\in S} \in \mathcal{O}^\varepsilon$. Since $\tilde{d}_i+\tilde{r}_i\boldsymbol{X}_S \sim_i d_i^k+r_i^k\boldsymbol{X}_N$ for all $i \in S$ it holds that $(\bar{d}^k,\bar{r}^k) \in W_S((d^k,r^k))$. Hence, $(\bar{d}^k,\bar{r}^k) \in PO_S^*((d^k,r^k))$.

Second, suppose that $(\tilde{d},\tilde{r}) \in PD_\Gamma(S)$ and that $\tilde{d}_i+\tilde{r}_i\boldsymbol{X}_S \prec_i \bar{d}_i+\bar{r}_i\boldsymbol{X}_S+\varepsilon$ for all $i \in S$. Since the Pareto optimality of $(\bar{d},\bar{r})$ implies that $(\bar{d}+\frac{1}{2}(\varepsilon,\varepsilon,\ldots,\varepsilon),\bar{r}) \in NPD_\Gamma(S)$ it follows from Proposition 5.5 that there exists $(\bar{d}^k,\bar{r}^k) \in PO_\Gamma(S)$ such that

$$\tilde{d}_i+\tilde{r}_i\boldsymbol{X}_S \precsim_i \bar{d}_i^k+\bar{r}_i^k\boldsymbol{X}_S \precsim_i \bar{d}_i+\bar{r}_i\boldsymbol{X}_S+\tfrac{1}{2}\varepsilon$$

for all $i \in S$. Thus, $(\bar{d}_i^k+\bar{r}_i^k\boldsymbol{X}_S)_{i\in S} \in \mathcal{O}^\varepsilon$. Since $\tilde{d}_i+\tilde{r}_i\boldsymbol{X}_S \sim_i d_i^k+r_i^k\boldsymbol{X}_N$ for all $i \in S$ it holds that $(\bar{d}^k,\bar{r}^k) \in B_S((d^k,r^k))$. Hence, $(\bar{d}^k,\bar{r}^k) \in PO_S^*((d^k,r^k))$.

Finally, suppose that $(\tilde{d},\tilde{r}) \in PD_\Gamma(S)$ and that $\tilde{d}_i+\tilde{r}_i\boldsymbol{X}_S \succ_i \bar{d}_i+\bar{r}_i\boldsymbol{X}_S-\varepsilon$ for all $i \in S$. Then Proposition 5.5 implies that there exists $(\bar{d}^k,\bar{r}^k) \in PO_\Gamma(S)$ such that

$$\bar{d}_i+\bar{r}_i\boldsymbol{X}_S-\varepsilon \precsim_i \tilde{d}_i+\tilde{r}_i\boldsymbol{X}_S \precsim_i \bar{d}_i^k+\bar{r}_i^k\boldsymbol{X}_S$$

for all $i \in S$. Thus, $(\bar{d}^k,\bar{r}^k) \in B_S((d^k,r^k))$. Therefore we have that $(\bar{d}^k,\bar{r}^k) \in PO_S^*((d^k,r^k))$. Moreover, $(\bar{d}_i^k+\bar{r}_i^k\boldsymbol{X}_S)_{i\in S} \in \mathcal{O}^\varepsilon$.

Now we are able to construct a sequence $(\sum_{i\in S}\delta_i^k)_{k\in\mathbb{N}}$ satisfying $\sum_{i\in S}\delta_i^k \in PO_S^*((d^k,r^k))$ for each $k \in \mathbb{N}$, such that each convergent subsequence converges to $\sum_{i\in S}\delta_i$.

Let $(\varepsilon^m)_{m\in\mathbb{N}}$ be a strictly decreasing sequence such that $\varepsilon^m > 0$ for all $m \in \mathbb{N}$ and $\lim_{m\to\infty}\varepsilon^m = 0$. Hence, $(\mathcal{O}^{\varepsilon^m})_{m\in\mathbb{N}}$ is a decreasing sequence in the sense that $\mathcal{O}^{\varepsilon^m} \subset \mathcal{O}^{\varepsilon^{m'}}$ if $m > m'$. Define $\mathcal{O}^0 = \cap_{\varepsilon>0}\mathcal{O}^\varepsilon$. From $(\bar{d},\bar{r}) \in PO_S^*((d,r))$ it follows that $(d_i+r_i\boldsymbol{X}_N)_{i\in S} \in \mathcal{O}^0$. Hence, $(d_i+r_i\boldsymbol{X}_N)_{i\in S} \in \mathcal{O}^\varepsilon$ for all $\varepsilon > 0$. Since (d^k,r^k) converges to (d,r) there exists $K^1 \in \mathbb{N}$ such that for all $k > K^1$ it holds that $(d_i^k+r_i^k\boldsymbol{X}_S)_{i\in S} \in \mathcal{O}^{\varepsilon^1}$. Next, take $k \in \mathbb{N}$. If $k \le K^1$ then take $\sum_{i\in S}\delta_i^k \in \overline{E}_S^2((d^k,r^k))$ arbitrary. If $k > K^1$ we distinguish the following two cases.

In the first case, suppose that $(d_i^k+r_i^k\boldsymbol{X}_N)_{i\in S} \in \mathcal{O}^0$. Then it holds true that $(\bar{d},\bar{r}) \in PO_S^*((d^k,r^k))$ and $(\bar{d}_i+\bar{r}_i\boldsymbol{X}_S)_{i\in S} \in \mathcal{O}^0$. So, we can take $(\bar{d}^k,\bar{r}^k)$ equal to $(\bar{d},\bar{r})$ (See Figure 5.10).

In the second case, let $(d_i^k+r_i^k\boldsymbol{X}_N)_{i\in S} \notin \mathcal{O}^0$. Then there exists $m(k) \in \mathbb{N}$ such that $(d_i^k+r_i^k\boldsymbol{X}_N)_{i\in S} \in \mathcal{O}^{\varepsilon^{m(k)}}\backslash\mathcal{O}^{\varepsilon^{m(k)+1}}$. Subsequently, take $(\bar{d}^k,\bar{r}^k) \in$

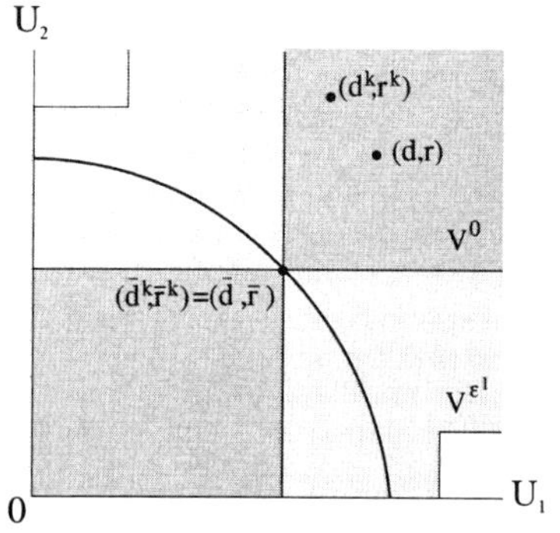

Figure 5.10

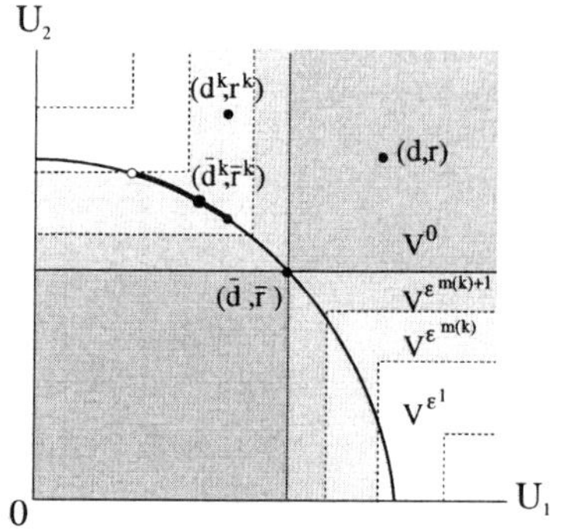

Figure 5.11

$PO^*_S((d^k,r^k))$ such that $(\bar{d}^k_i+\bar{r}^k_i\boldsymbol{X}_S)_{i\in S}\in\mathcal{O}^{\varepsilon^{m(k)}}$ (See Figure 5.11, where the bold printed curve represents the set of allocations that belong to both $PO^*_S((d^k,r^k))$ and $\mathcal{O}^{\varepsilon^{m(k)}}$). That such $(\bar{d}^k,\bar{r}^k)$ exists can be seen as follows. Let $(d',r')\in I_\Gamma(S)$ be such that $d'_i+r'_i\boldsymbol{X}_S\sim_i d^k_i+r^k_i\boldsymbol{X}_N$ for all $i\in S$. Then either $(d',r')\in PD_\Gamma(S)$ or $(d',r')\in NPD_\Gamma(S)$.

For the case that $(d',r')\in PD_\Gamma(S)$ then $(d^k_i+r^k_i\boldsymbol{X}_N)_{i\in S}\in\mathcal{O}^{\varepsilon^{m(k)}}$ implies that $(d'_i+r'_i\boldsymbol{X}_S)_{i\in S}\in\mathcal{O}^{\varepsilon^{m(k)}}$ and, consequently, that

$$d'_i+r'_i\boldsymbol{X}_S\succsim_i\bar{d}_i+\bar{r}_i\boldsymbol{X}_S-\varepsilon^{m(k)}$$

for all $i\in S$ or

$$d'_i+r'_i\boldsymbol{X}_S\precsim_i\bar{d}_i+\bar{r}_i\boldsymbol{X}_S+\varepsilon^{m(k)}$$

for all $i\in S$. If the first statement is true then it follows from Proposition 5.5 that there exists $(\bar{d}^k,\bar{r}^k)\in PO_\Gamma(S)$ such that $\bar{d}^k_i+\bar{r}^k_i\boldsymbol{X}_S\succsim_i d'_i+r'_i\boldsymbol{X}_S$ for all $i\in S$. This implies that $(d^k_i+\bar{r}^k_i\boldsymbol{X}_S)_{i\in S}\in\mathcal{O}^{\varepsilon^{m(k)}}$ and $(\bar{d}^k,r^k)\in PO^*_S((d^k,r^k))$. If the second statement is true then it follows from $(d',r')\in PD_\Gamma(S)$ and Proposition 5.5 that there exists $(\bar{d}^k,\bar{r}^k)\in PO_\Gamma(S)$ satisfying

$$d^k_i+r^k_i\boldsymbol{X}_N\sim_i d'_i+r'_i\boldsymbol{X}_S\precsim_i\bar{d}^k_i+\bar{r}^k_i\boldsymbol{X}_S\precsim_i\bar{d}_i+\bar{r}_i\boldsymbol{X}_S+\varepsilon^{m(k)}$$

for all $i\in S$. Hence, $(\bar{d}^k,\bar{r}^k)\in PO^*_S((d^k,r^k))$ and $(\bar{d}^k_i+\bar{r}^k_i\boldsymbol{X}_S)_{i\in S}\in\mathcal{O}^{\varepsilon^{m(k)}}$.

For the case that $(d',r')\in NPD_\Gamma(S)$ a similar argument holds.

Next, let δ^k_i be such that $\bar{d}^k_i+\bar{r}^k_i\boldsymbol{X}_S\sim_i d^k_i+r^k_i\boldsymbol{X}_N+\delta^k_i$ for all $i\in S$. Note that for $k>K^1$ we have $\sum_{i\in S}\delta^k_i\in\overline{E}^2_S((d^k,r^k))$ and $(\bar{d}^k_i+\bar{r}^k_i\boldsymbol{X}_S)_{i\in S}\in\mathcal{O}^0$ if $(d^k_i+r^k_i\boldsymbol{X}_N)_{i\in S}\in\mathcal{O}^0$ and $(\bar{d}^k_i+\bar{r}^k_i\boldsymbol{X}_S)_{i\in S}\in\mathcal{O}^{\varepsilon^{m(k)}}$ if $(d^k_i+r^k_i\boldsymbol{X}_N)_{i\in S}\notin\mathcal{O}^0$. Since $(\sum_{i\in S}\delta^k_i)_{k\in\mathbb{N}}$ is a sequence in the compact set $\overline{E}^2_S(IR_\Gamma(S))$ there exists a convergent subsequence $(\sum_{i\in S}\delta^l_i)_{l\in\mathbb{N}}$ with limit $\sum_{i\in S}\tilde{\delta}_i$. Corresponding to this convergent subsequence there is a sequence $(\bar{d}_i^l + \bar{r}_i^l \boldsymbol{X}_S)_{l \in \mathbb{N}}$ such that $(\bar{d}_i^l + \bar{r}_i^l \boldsymbol{X}_S)_{i \in S} \in \mathcal{O}^0$ if $(d_i^l + r_i^l \boldsymbol{X}_N)_{i \in S} \in \mathcal{O}^0$ and $(\bar{d}_i^l + \bar{r}_i^l \boldsymbol{X}_S)_{i \in S} \in \mathcal{O}^{\varepsilon^{m(l)}}$ if $(d_i^l + r_i^l \boldsymbol{X}_N)_{i \in S} \notin \mathcal{O}^0$. Moreover, it holds that $\bar{d}_i^l + \bar{r}_i^l \boldsymbol{X}_S \sim_i d_i^l + r_i^l \boldsymbol{X}_N + \delta_i^l$ for all $i \in S$. This implies that $(d_i^l + r_i^l \boldsymbol{X}_N + \delta_i^l)_{i \in S} \in \mathcal{O}^0$ if $(d_i^l + r_i^l \boldsymbol{X}_N)_{i \in S} \in \mathcal{O}^0$ and $(d_i^l + r_i^l \boldsymbol{X}_N + \delta_i^l)_{i \in S} \in \mathcal{O}^{\varepsilon^{m(l)}}$ if $(d_i^l + r_i^l \boldsymbol{X}_N)_{i \in S} \notin \mathcal{O}^0$. From $\mathcal{O}^0 = \cap_{l \in \mathbb{N}} \mathcal{O}^{\varepsilon^{m(l)}}$ it follows that

$$\lim_{l \to \infty} (d_i^l + r_i^l \boldsymbol{X}_N + \delta_i^l)_{i \in S} = (d_i + r_i \boldsymbol{X}_N + \tilde{\delta}_i)_{i \in S} \in \mathcal{O}^0.$$

This implies that $d_i + r_i \boldsymbol{X}_N + \tilde{\delta}_i \sim_i \bar{d}_i + \bar{r}_i \boldsymbol{X}_S$ for all $i \in S$. Since $\bar{d}_i + \bar{r}_i \boldsymbol{X}_S \sim_i d_i + r_i \boldsymbol{X}_N + \delta_i$ for all $i \in S$ assumption (C2) implies that $\tilde{\delta}_i = \delta_i$ for all $i \in S$. Consequently, it holds that

$$\lim_{l \to \infty} \sum_{i \in S} \delta_i^l = \sum_{i \in S} \tilde{\delta}_i = \sum_{i \in S} \delta_i.$$

So, each convergent subsequence $(\sum_{i \in S} \delta_i^l)_{l \in \mathbb{N}}$ converges to $\sum_{i \in S} \delta_i$. Hence, $(\sum_{i \in S} \delta_i^k)_{k \in \mathbb{N}}$ converges to $\sum_{i \in S} \delta_i$, which completes the proof. □

Lemma 5.16 The excess function $E_S((d,r))$ is continuous in (d,r) for each $(d,r) \in IR_\Gamma(S)$.

PROOF: Let $((d^k, r^k))_{k \in \mathbb{N}}$ be a sequence in $IR_\Gamma(N)$ converging to $(d,r) \in IR_\Gamma(N)$. We have to show that $\lim_{k \to \infty} E_S((d^k, r^k)) = E_S((d,r))$. Since $(E_S((d^k, r^k)))_{k \in \mathbb{N}}$ is a sequence in the compact set $\overline{E}_S^2(IR_\Gamma(N))$ there exists a convergent subsequence $(E_S((d^l, r^l)))_{l \in \mathbb{N}}$ with limit η. Note that the upper semi continuity of $\overline{E}_S^2$ implies that $\eta \in \overline{E}_S^2((d,r))$. Since $\overline{E}_S^2$ is lower semi continuous there exists a sequence $(\sum_{i \in S} \delta_i^l)_{l \in \mathbb{N}}$ such that $\sum_{i \in S} \delta_i^l \in \overline{E}_S^2((d^l, r^l))$ for all $l \in \mathbb{N}$ and $\lim_{l \to \infty} \sum_{i \in S} \delta_i^l = E_S((d,r))$. Then

$$\begin{array}{lllll} \lim_{l \to \infty} E_S((d^l, r^l)) & \leq & \lim_{l \to \infty} \sum_{i \in S} \delta_i^l & = & E_S((d,r)) \\ & \leq & \eta & = & \lim_{l \to \infty} E_S((d^l, r^l)). \end{array}$$

Hence, $\lim_{l \to \infty} \sum_{i \in S} E_S((d^l, r^l)) = E_S((d,r))$. Thus, every convergent subsequence of $(E_S((d^k, r^k)))_{k \in \mathbb{N}}$ converges to $E_S((d,r))$. The compactness of $\overline{E}_S^2(IR_\Gamma(N))$ then implies that $\lim_{k \to \infty} E_S((d^k, r^k)) = E_S((d,r))$. □

6 Risk Sharing and Insurance

Within the bounds of their possibilities, individuals generally try to eliminate the risks resulting from their social and economic activities as much as possible. In this regard, a common way of doing so is by means of risk sharing and/or insurance. Firms, for instance, might prefer to cooperate in joint ventures when investing in risky projects, while investors perpetrate risk sharing by investing their capital in a diversified portfolio of financial assets.

In insurance, an insurance company contracts to bear (part of) an individual's risk in exchange for a fixed payment, the insurance premium. For actuarial scientists, this insurance premium has been and still is a much examined research topic. The main question they address is, what is a reasonable premium for the risk that is insured, for the insurance premium should be acceptable with respect to the opposite interests of both the insurer and the insured.

Classical actuarial theory mainly considers this problem from the insurer's point of view. In determining a reasonable insurance premium, it distinguishes between risk arising from the 'life' sector and risk arising from the 'non life' sector. For the first, there is a profusion of statistical data on the expected remaining life available, which makes the calculation of an appropriate premium relatively easy. For the latter, however, things are a bit more complicated. In 'non life' insurance the risk is not always easy to capture in a statistical framework. Therefore, several premium calculation principles have been developed to serve this purpose, see for instance Goovaerts, De Vylder and Haezendonck (1984).

These calculation principles, however, only take into account a part of the insurer's side of the deal. More precisely, they consider whether the premium is high enough to cover the risk. Competition arising from the presence of other insurers on the one hand, and the interests of the insured, on the other hand, are mostly ignored. It is, of course, better to consider all these aspects in an insurance deal, since the premium should not only be high enough to compensate the insurer for bearing the individual's risk, it should also be low enough so that an individual is willing to insure his risk (or a part of it) for this premium. The economic models for (re)insurance markets, which were developed from the 1960's on (cf. Borch (1962a) and Bühlmann (1980), (1984)), consider indeed the interests of both the insurers and the insureds. These models incorporate the possibility to study problems concerning fairness, Pareto optimality and market equilibrium. Bühlmann (1980), for example, shows that the Esscher calculation principle results in a Pareto optimal outcome.

More recently, also game theory is used to model the interests of all parties in an insurance problem. Examples can be found in Borch (1962b), Lemaire (1991), and Alegre and Mercè Claramunt (1995), and Suijs, De Waegenaere and Borm (1998). The latter uses stochastic cooperative games to model individual insurance as well as reinsurance by insurance companies. The results they obtain, however, only hold for exponentially distributed losses. This chapter generalizes Suijs et al. (1998) in the sense that it allows arbitrary random losses. We determine Pareto optimal allocations and show that the zero utility principle for calculating premiums (see Goovaerts et al. (1984)) yields a core allocation.

6.1 Insurance Games

For modeling insurance problems we use a slightly modified version of our model of a stochastic cooperative game as introduced in Chapter 3. We show that by cooperating, individuals and insurers can redistribute their risks and, consequently, improve their welfare. First, we need to specify the agents that participate in the game. An agent can be one of two types; an agent is either an individual person or an insurer. The set of individual persons is denoted by N_P and the set of insurers is denoted by N_I. Hence, the agents of the game are denoted by the set $N_I \cup N_P$.

Again, all agents are assumed to be risk averse expected utility maximizers with utility function $U_i(t) = \beta_i e^{-\alpha_i t}$, $(t \in \mathbb{R})$, with $\beta_i < 0$, $\alpha_i > 0$ for all $i \in N_I \cup N_P$. By changing the signs of the parameters β_i and α_i the utility function becomes convex, and, as a consequence, the agent will be risk loving. Regarding the situations where one or more risk neutral/loving insurers are involved we confine ourselves to a brief discussion later on.

Next, let $-\boldsymbol{X}_i$ with $\boldsymbol{X}_i \in L^1(\mathbb{R}_+)$ describe the future random losses of agent $i \in N_I \cup N_P$. For an individual $i \in N_P$, the variable $-\boldsymbol{X}_i$ describes the random losses that could occur to this individual. They include, for example, the monetary damages caused by cars, bikes, fires, or other people. For an insurance company $i \in N_I$, the variable $-\boldsymbol{X}_i$ describes the random losses corresponding to its insurance portfolio. We assume that the random losses $-\boldsymbol{X}_i$, $(i \in N_I \cup N_P)$, are mutually independent.

Now, let us focus on the possibilities that occur when agents decide to cooperate. Therefore, consider a coalition S of agents. If the members of S decide to cooperate, the total loss $\boldsymbol{X}_S \in L^1(\mathbb{R})$ of coalition S equals the sum of the individual losses of the members of S, i.e., $\boldsymbol{X}_S = -\sum_{i \in S} \boldsymbol{X}_i$. Subsequently, the loss $\boldsymbol{X}_S$ has to be allocated to the members of S.

In Chapter 3 we described an allocation of the random payoff $\boldsymbol{X}_S$ to the members of coalition S by means of a pair (d, r). Applying this definition to insurance games, however, raises a problem. Given an allocation (d, r) we have that agent $i \in S \cap N_P$ receives $d_i + r_i \boldsymbol{X}_S$. Thus $\boldsymbol{X}_S$ not only consists of the future random losses of agent i, but also of the future random losses of all other individuals $j \in S \cap N_P$. Hence, if agent i receives $d_i + r_i \boldsymbol{X}_S$ he receives (part of) the random losses of his fellow agents $j \in S \cap N_P$. Furthermore, this means that an agent $j \in S \cap N_P$ transfers (part of) his random losses to agent i, or, put in other words, agent j insures (part of) his random losses at agent i. But this is rather unusual; agents only make insurance deals with insurance companies and not with other individuals. So, we need to modify our definition of an allocation so as to incorporate transfers of random losses from individuals to insurance companies only. The option we choose for is to replace the vector $r \in \Delta^S$ by a matrix $R \in \mathbb{R}^{S \times S}$. An element r_{ij} then represents the fraction of agent j's random loss that he transfers to agent i. Then by imposing the right conditions on R we can guarantee that individuals cannot transfer any risks among each other.

For explaining an allocation of the loss $\boldsymbol{X}_S$ in more detail, we distinguish between the following three cases. In the first case, coalition S consists of insurers only. So, $S \subset N_I$. Such a coalition is assumed to allocate the loss $\boldsymbol{X}_S$ in the following way. First, a coalition S allocates a fraction $r_{ij} \in [0, 1]$ of the loss $\boldsymbol{X}_j$ of insurer $j \in S$ to insurer $i \in S$. So, insurer i bears a total loss of $\sum_{j \in S} r_{ij} \boldsymbol{X}_j$, where $r_{ij} \in [0, 1]$ and $\sum_{k \in S} r_{kj} = 1$. This is called proportional (re)insurance. This part of the allocation of $\boldsymbol{X}_S$ for coalition S is described by a matrix $R \in \mathbb{R}_+^{S \times S}$, where r_{ij} represents the fraction insurer i bears of insurer j's loss $\boldsymbol{X}_j$. Second, the insurers are allowed to make deterministic transfer payments. This means that each insurance company $i \in S$ also receives an amount $d_i \in \mathbb{R}$ such that $\sum_{j \in S} d_j \leq 0$. These transfer payments can be interpreted as the aggregate premium insurers have to pay for the actual risk exchanges.

In the second case, coalition S consists of individual persons only. So, $S \subset N_P$. Then the gains of cooperation are assumed to be nil. That is, we do not allow any risk exchanges between the persons themselves. For, that is what the insurers are for in the first place. As a result, the only allocations (d, R) of $\boldsymbol{X}_S$ which are allowed are of the form $r_{ii} = 1$ for all $i \in S$ and $r_{ij} = 0$ for all $i, j \in S$ with $i \neq j$.

In the third and last case, coalition S consists of both insurers and individual persons. So, $S \subset N_I \cup N_P$. Now cooperation can take place in two different ways. First, insurers are allowed to exchange (parts of) their portfolios with other insurers, and, second, individual persons may transfer (parts of) their risks to insurers. Again, individual persons are not allowed to exchange risks with each other. Moreover, we assume that insurers cannot transfer (parts of) their portfolios to individuals.

Summarizing we can say that there are several restrictions on allocations. To be more precise, denote by S_I the set of insurers of coalition S, i.e., $S_I = S \cap N_I$, and by S_P the set of individuals of coalition S, i.e., $S_P = S \cap N_P$. Then an allocation $(d, R) \in \mathbb{R}^S \times \mathbb{R}_+^{S \times S}$ is feasible for the coalition S if for all $i \in S_P$ and all $j \in S$ with $i \neq j$ it holds that $r_{ij} = 0$ and $\sum_{i \in S} r_{ij} = 1$ for all $j \in S$. Furthermore, given an allocation (d, R) of the random loss $\boldsymbol{X}_S$, agent $i \in S$ receives $d_i + \sum_{j \in S} r_{ij} X_j = d_i + R_i \boldsymbol{X}^S$, where R_i denotes the i-th row of R and $\boldsymbol{X}^S = (-\boldsymbol{X}_i)_{i \in S}$.

Example 6.1 Let $N_I = \{1, 2\}$, $N_P = \{3, 4\}$ and consider the coalition $S = \{1, 3, 4\}$. Then $\boldsymbol{X}_S = -\boldsymbol{X}_1 - \boldsymbol{X}_3 - \boldsymbol{X}_4$. A feasible allocation for S is the following. Let $d = (3, -2, -1)$ and $r_{11} = 1$, $r_{13} = \frac{1}{2}$, $r_{33} = \frac{1}{2}$, $r_{14} = \frac{1}{5}$ and $r_{44} = \frac{4}{5}$. Then insurer 1 receives

$$d_1 + R_1 \boldsymbol{X}^S \quad = \quad 3 - \boldsymbol{X}_1 - \tfrac{1}{2}\boldsymbol{X}_3 - \tfrac{1}{5}\boldsymbol{X}_4),$$

individual 3 receives

$$d_3 + R_3 \boldsymbol{X}^S \quad = \quad -2 - \tfrac{1}{2}\boldsymbol{X}_3,$$

and individual 4 receives

$$d_4 + R_4 \boldsymbol{X}^S \quad = \quad -1 - \tfrac{4}{5}\boldsymbol{X}_4.$$

So, individuals 3 and 4 pay a premium of 2 and 1, respectively, to insurer 1 for the insurance of their losses.

In conclusion, an *insurance game* Γ with agent set $N_I \cup N_P$ is described by the tuple $(N_I \cup N_P, \mathcal{V}, (U_i)_{i \in N_I \cup N_P})$, where N_I is the set of insurers, N_P the set of individuals, $\mathcal{V}(S) = \{\sum_{i \in S} -\boldsymbol{X}_i\}$ the random loss for coalition S, and U_i the utility function for agent $i \in N_I \cup N_P$. The class of all such insurance games with insurers N_I and individuals N_P is denoted by $IG(N_I, N_P)$.

6.1.1 Pareto Optimal Distributions of Risk

Since the preferences of each agent are described by means of a utility function, we can confine ourselves to considering certainty equivalents (see Section 3.2.2). In this model, the certainty equivalent is defined by $m_i(\boldsymbol{X}) = U_i^{-1}(E(U_i(\boldsymbol{X})))$

provided that the expected utility exists.[1] Since the exponential utility functions satisfy conditions (M1) and (M2), the results stated in Section 3.2.2 on certainty equivalents apply. One of these results concerns the Pareto optimality of an allocation. For insurance games this result reads as follows.

Proposition 6.2 Let $\Gamma \in IG(N_I, N_P)$ and $S \subset N_I \cup N_P$. An allocation $(d_i + R_i\boldsymbol{X}^S)_{i\in S} \in \mathcal{Z}_\Gamma(S)$ is Pareto optimal for coalition S if and only if

$$\begin{aligned}\sum_{i\in S} m_i(d_i + R_i\boldsymbol{X}^S) \\ = \max\left\{\sum_{i\in S} m_i(\tilde{d}_i + \tilde{R}_i\boldsymbol{X}^S) \mid (\tilde{d}_i + \tilde{R}_i\boldsymbol{X}^S)_{i\in S} \in \mathcal{Z}_\Gamma(S)\right\}.\end{aligned} \tag{6.1}$$

So, an allocation is Pareto optimal for coalition S if and only if this allocation maximizes the sum of the certainty equivalents. To determine these allocations, we first need to calculate the certainty equivalent of an allocation $(d_i + R_i\boldsymbol{X}^S)_{i\in S} \in \mathcal{Z}_\Gamma(S)$ for agent $i \in S$. Therefore, let $S \subset N_I \cup N_P$ and $(d_i + R_i\boldsymbol{X}^S)_{i\in S} \in \mathcal{Z}_\Gamma(S)$. The random loss coalition S has to allocate equals $\boldsymbol{X}_S = \sum_{i\in S}\boldsymbol{X}_i$. Given a feasible allocation $(d_i + R_i\boldsymbol{X}^S)_{i\in S} \in \mathcal{Z}_\Gamma(S)$, the random payoff to agent $i \in S$ equals

$$d_i + R_i\boldsymbol{X}^S = d_i - \sum_{j\in S} r_{ij}\boldsymbol{X}_j$$

if $i \in S_I$ and

$$d_i + R_i\boldsymbol{X}^S = d_i - r_{ii}\boldsymbol{X}_i$$

if $i \in S_P$. Consequently, we have that the certainty equivalent of $(d_i + R_i\boldsymbol{X}^S)_{i\in S}$ equals

$$m_i(d_i + R_i\boldsymbol{X}^S) = \begin{cases} d_i + \frac{1}{\alpha_i}\log\left(\int_0^\infty e^{\alpha_i r_{ii} t} dF_{\boldsymbol{X}_i}(t)\right)^{-1}, & \text{if } i \in S_P, \\ d_i + \sum_{j\in S}\frac{1}{\alpha_i}\log\left(\int_0^\infty e^{\alpha_i r_{ij} t} dF_{\boldsymbol{X}_j}(t)\right)^{-1}, & \text{if } i \in S_I. \end{cases} \tag{6.2}$$

The sum of the certainty equivalents then equals

$$\begin{aligned}\sum_{i\in S} m_i(d_i + R_i\boldsymbol{X}^S) &= \sum_{i\in S} d_i + \sum_{i\in S_P}\frac{1}{\alpha_i}\log\left(\int_0^\infty e^{\alpha_i r_{ii} t} dF_{\boldsymbol{X}_i}(t)\right)^{-1} \\ &\quad + \sum_{i\in S_I}\sum_{j\in S}\frac{1}{\alpha_i}\log\left(\int_0^\infty e^{\alpha_i r_{ij} t} dF_{\boldsymbol{X}_j}(t)\right)^{-1}.\end{aligned} \tag{6.3}$$

[1] Throughout this chapter we assume that the utility functions and random payoffs are such that the expected utility exist, and that we may interchange the order of integration and differentiation.

Since $\sum_{i\in S} d_i = 0$ for Pareto optimal allocations, we have for these allocations that the sum of the certainty equivalents is independent of the vector of transfer payments d. Intuitively, this is quite clear. For since $\sum_{h\in S} d_h \leq 0$, an increase in d_i for agent i implies that d_j decreases for at least one other agent j. Consequently, Pareto optimality is solely determined by the choice of the allocation risk exchange matrix R of the random losses. In fact, the next theorem shows that there is a unique allocation risk exchange matrix R^* inducing Pareto optimality.

Theorem 6.3 Let $\Gamma \in IG(N_I, N_P)$ and $S \subset N_I \cup N_P$. An allocation $(d_i + R_i^* \boldsymbol{X}^S)_{i\in S} \in \mathcal{Z}_\Gamma(S)$ is Pareto optimal for S if and only if

$$r_{ij}^* = \begin{cases} \dfrac{\frac{1}{\alpha_i}}{\sum_{h\in S_I} \frac{1}{\alpha_h}} & \text{, if } i,j \in S_I, \\ \dfrac{\frac{1}{\alpha_i}}{\sum_{h\in S_I\cup\{j\}} \frac{1}{\alpha_h}} & \text{, if } i \in S_I \cup \{j\} \text{ and } j \in S_P, \\ 0 & \text{, otherwise.} \end{cases}$$

PROOF: We have to show that R^* is the unique solution of

$$\begin{aligned} \max \sum_{i\in S_P} \tfrac{1}{\alpha_i} \log \left(\int_0^\infty e^{\alpha_i r_{ii} t} dF_{\boldsymbol{X}_i}(t) \right)^{-1} \\ + \sum_{i\in S_I} \sum_{j\in S} \tfrac{1}{\alpha_i} \log \left(\int_0^\infty e^{\alpha_i r_{ij} t} dF_{\boldsymbol{X}_j}(t) \right)^{-1} \end{aligned}$$

$$\begin{aligned} \text{s.t.:} \quad r_{jj} + \textstyle\sum_{i\in S_I} r_{ij} &= 1, \quad \text{for all } j \in S_P, \\ \textstyle\sum_{i\in S_I} r_{ij} &= 1, \quad \text{for all } j \in S_I, \\ r_{ii} &\geq 0, \quad \text{if } i \in S_P, \\ r_{ij} &\geq 0, \quad \text{if } i \in S_I \text{ and } j \in S. \end{aligned}$$

Lemma 6.6 with $c = \alpha_i$ and $x = r_{ij}$ for all relevant combinations of $i, j \in S$, implies that the objective function is strictly concave. Hence, it is sufficient to prove that R^* is a solution of this maximization problem. The Karush-Kuhn-Tucker conditions[2] tell us that this is indeed the case if there exists $\lambda_j \in \mathbb{R}$ $(j \in S)$,

[2] The Karush-Kuhn-Tucker conditions read as follows:

$$\begin{aligned} \text{If } f(x) \;=\; & \max_y f(y) \\ & \text{s.t. } g_k(y) \leq 0, \quad k \in K \\ & \phantom{\text{s.t. }} g_l(y) = 0, \quad l \in L \end{aligned}$$

then there exist $\nu_k \geq 0$ $(\forall k \in K)$ and $\lambda_l \in \mathbb{R}$ $(\forall l \in L)$ such that

$$\begin{aligned} \nabla f(x) &= \textstyle\sum_{k\in K} \nu_k \cdot \nabla g_k(x) + \sum_{l\in L} \lambda_l \cdot \nabla g_l(x) \\ \nu_k \cdot g_k(x) &= 0, \text{ for all } k \in K. \end{aligned}$$

Moreover, if f is strictly concave and g_k $(k \in K)$, g_l $(l \in L)$ are convex then the reverse of the statement also holds and the maximum is unique.

$\nu_{jj} \geq 0$ $(j \in S_P)$ and $\nu_{ij} \geq 0$ $(i \in S_I,\ j \in S)$ such that

$$\begin{aligned}
-\frac{\int_0^\infty te^{\alpha_j r^*_{jj} t} dF_{\boldsymbol{X}_j}(t)}{\int_0^\infty e^{\alpha_j r^*_{jj} t} dF_{\boldsymbol{X}_j}(t)} &= \lambda_j - \nu_{jj}, \quad \text{for all } j \in S_P, \\
-\frac{\int_0^\infty te^{\alpha_i r^*_{ij} t} dF_{\boldsymbol{X}_j}(t)}{\int_0^\infty e^{\alpha_i r^*_{ij} t} dF_{\boldsymbol{X}_j}(t)} &= \lambda_j - \nu_{ij}, \quad \text{for all } i \in S_I \text{ and all } j \in S, \\
\nu_{jj} r_{jj} &= 0, \quad \text{for all } j \in S_P, \\
\nu_{ij} r_{ij} &= 0, \quad \text{for all } i \in S_I \text{ and all } j \in S.
\end{aligned}$$

Substituting r^*_{ij} gives $\nu_{ij} = 0$ for all relevant combinations of $i, j \in S$ and

$$\begin{aligned}
\lambda_j &= -\frac{\int_0^\infty te^{t/\alpha(S_I \cup \{j\})} dF_{\boldsymbol{X}_j}(t)}{\int_0^\infty e^{t/\alpha(S_I \cup \{j\})} dF_{\boldsymbol{X}_j}(t)}, \quad \text{for all } j \in S_P, \\
\lambda_j &= -\frac{\int_0^\infty te^{t/\alpha(S_I)} dF_{\boldsymbol{X}_j}(t)}{\int_0^\infty e^{t/\alpha(S_I)} dF_{\boldsymbol{X}_j}(t)}, \quad \text{for all } i \in S_I,
\end{aligned}$$

with $\alpha(S) = \sum_{i \in S} \frac{1}{\alpha_i}$. Consequently, R^* is the optimal solution. □

So, for a Pareto optimal allocation of a loss $\boldsymbol{X}_j$ within S one has to distinguish between two cases. In the first case the index j refers to an insurer and in the second case j refers to an individual. When $\boldsymbol{X}_j$ is the loss of insurer $j \in S_I$, the loss is allocated proportionally to $\frac{1}{\alpha_i}$ among all insurers in coalition S. When $\boldsymbol{X}_j$ is the loss of individual $j \in S_P$, the loss is allocated proportionally to $\frac{1}{\alpha_i}$ among all insurers in coalition S and individual j himself. Note that by the feasibility constraints nothing is allocated to the other individuals. Furthermore, remark that if only reinsurance of the insurance portfolios is considered, that is, $N_P = \emptyset$ then the Pareto optimal allocation coincides with the Pareto optimal allocation of (re)insurance markets discussed in Bühlmann (1980).

The determination of the allocation risk exchange matrix is, of course, only one part of the allocation. We still have to determine the vector of transfer payments d, that is, the premiums that have to be paid. Although an allocation $(d_i + R^*_i \boldsymbol{X}^S)_{i \in S}$ may be Pareto optimal for any choice of d, not every d is satisfactory from a social point of view. An insurer will not agree with insuring the losses of other agents if he is not properly compensated, that is, if he does not receive a fair premium for the insurance. Similarly, insurance companies and individuals only agree to insure their losses if the premium they have to pay is reasonable. Consequently, there is a conflict of interests; both insurance companies and individuals want to pay a low premium for insuring their own losses, while insurance companies want to receive a high premium for bearing the losses of other agents.

6.1.2 The Zero Utility Premium Calculation Principle

Premium calculation principles indicate how to determine the premium for a certain risk. In the past, various of these principles were designed, for example, the net premium principle, the expected value principle, the standard deviation principle, the Esscher principle, and the zero utility principle (cf. Goovaerts et al. (1984). In this section we focus on the zero utility principle. A premium calculation principle determines a premium $\pi_i(\boldsymbol{X})$ for individual i for bearing the risk $\boldsymbol{X}$. The zero utility principle assigns a premium $\pi_i(\boldsymbol{X})$ to $\boldsymbol{X}$ such that the utility level of individual i, who bears the risk $\boldsymbol{X}$, remains unchanged when the wealth w_i of this individual changes to $w_i + \pi_i(\boldsymbol{X}) - \boldsymbol{X}$. Since individuals are expected utility maximizers this means that the premium $\pi_i(\boldsymbol{X})$ satisfies $U_i(w_i) = E(U_i(w_i + \pi_i(\boldsymbol{X}) - \boldsymbol{X}))$. Note that the premium of the risk $\boldsymbol{X}$ depends on the individual who bears this risk and his wealth w_i.

Now, let us return to insurance games and utilize the zero utility principle to determine the allocation transfer payments $d \in \mathbb{R}^{N_I \cup N_P}$. At first this might seem difficult since the zero utility principle requires initial wealths w_i which do not appear in our model of insurance games. The exponential utility functions, however, yield that the zero utility principle is independent of these initial wealths w_i. To see this, let $\Gamma \in IG(N_I, N_P)$ be an insurance game. Since utility functions are exponential we can rewrite the expression $U_i(w_i) = E(U_i(w_i + \pi_i(\boldsymbol{X}) - \boldsymbol{X}))$ as follows

$$\begin{aligned} w_i &= U_i^{-1}(E(U_i(w_i + \pi_i(\boldsymbol{X}) - \boldsymbol{X}))) \\ &= w_i + \pi_i(\boldsymbol{X}) + U_i^{-1}(E(U_i(-\boldsymbol{X}))). \end{aligned}$$

Hence, $\pi_i(\boldsymbol{X}) = -U_i^{-1}(E(U_i(-\boldsymbol{X}))) = -m_i(-\boldsymbol{X})$ which indeed is independent of the wealth w_i. Furthermore, we can calculate the premium that agents receive for the risk they bear. For this, recall that for the Pareto optimal allocation risk exchange matrix R^* we have that

$$r^*_{ij} = \begin{cases} \dfrac{\frac{1}{\alpha_i}}{\sum_{h \in S_I} \frac{1}{\alpha_h}} & \text{, if } i, j \in S_I, \\ \dfrac{\frac{1}{\alpha_i}}{\sum_{h \in S_I \cup \{j\}} \frac{1}{\alpha_h}} & \text{, if } i \in S_I \cup \{j\} \text{ and } j \in S_P, \\ 0 & \text{, otherwise.} \end{cases}$$

Since the risk that insurer i bears equals $\sum_{j \in N_I \cup N_P} r^*_{ij} \boldsymbol{X}_j$, the premium he should receive for this according to the zero utility principle equals

$$\pi_i(\sum_{j \in N_I \cup N_P} r^*_{ij} \boldsymbol{X}_j) = -m_i(-\sum_{j \in N_I \cup N_P} r^*_{ij} \boldsymbol{X}_j)$$

Note that due to the mutual independence of $(\boldsymbol{X}_i)_{i\in N_I\cup N_P}$, the zero utility principle satisfies additivity, that is, $\pi_i(\sum_{j\in N_I\cup N_P} r^*_{ij}\boldsymbol{X}_j) = \sum_{j\in N_I\cup N_P} \pi_i(r^*_{ij}\boldsymbol{X}_j)$. As a consequence, we let the premium that individual $i \in N_P$ has to pay for insuring his loss at insurer j equal the zero utility premium that this insurer wants to receive for bearing this risk. Hence, individual i pays insurer j an amount $\pi_j(r^*_{ji}\boldsymbol{X}_i) = -m_j(-r^*_{ji}\boldsymbol{X}_i)$. Because individuals are not allowed to bear (part of) the risk of any other individual/insurer he does not receive any premium. So in aggregate he pays

$$\sum_{j\in N_I} \pi_j(r^*_{ji}\boldsymbol{X}_i).$$

Similarly, the premium that insurer i has to pay for reinsuring the fraction r^*_{ji} of his own portfolio $\boldsymbol{X}_i$ at insurer j, equals the premium that insurer j wants to receive for bearing this risk, that is, $\pi_j(r^*_{ji}\boldsymbol{X}_i) = -m_j(-r^*_{ji}\boldsymbol{X}_i)$. Then the premium insurer i receives in aggregate equals

$$\sum_{j\in N_I\cup N_P} \pi_i(r^*_{ij}\boldsymbol{X}_j) - \sum_{j\in N_I} \pi_j(r^*_{ji}\boldsymbol{X}_i).$$

Since

$$\sum_{i\in N_I}\left(\sum_{j\in N_I\cup N_P} \pi_i(r^*_{ij}\boldsymbol{X}_j) - \sum_{j\in N_I} \pi_j(r^*_{ji}\boldsymbol{X}_i)\right) - \sum_{i\in N_P}\sum_{j\in N_I} \pi_j(r^*_{ji}\boldsymbol{X}_i) = 0,$$

the zero utility principle yields an allocation transfer payments vector d^0 where

$$\begin{aligned} d^0_i &= \sum_{j\in N_I\cup N_P} \pi_i(r^*_{ij}\boldsymbol{X}_j) - \sum_{j\in N_I} \pi_j(r^*_{ji}\boldsymbol{X}_i) \\ &= -\sum_{j\in N_I\cup N_P} m_i(-r^*_{ij}\boldsymbol{X}_j) + \sum_{j\in N_I} m_j(-r^*_{ji}\boldsymbol{X}_i) \end{aligned} \tag{6.4}$$

for all $i \in N_I$ and

$$d^0_i = -\sum_{j\in N_I} \pi_j(r^*_{ji}\boldsymbol{X}_i) = \sum_{j\in N_I} m_j(-r^*_{ji}\boldsymbol{X}_i) \tag{6.5}$$

for all $i \in N_P$.

Theorem 6.4 Let $\Gamma \in IG(N_I, N_P)$. If d^0 is the vector of transfer payments determined by the zero utility premium calculation principle and R^* is the Pareto optimal risk exchange matrix then $(d^0_i + R^*_i\boldsymbol{X}^N)_{i\in N} \in \mathcal{C}(\Gamma)$.

PROOF: By Theorem 4.1 it suffices to show that $(m_i(d^0_i - R^*_i\boldsymbol{X}^N))_{i\in N} \in C(v_\Gamma)$. Hence, we must show that $\sum_{i\in S} m_i(d^0_i - R^*_i\boldsymbol{X}^N) \geq v_\Gamma(S)$ for all $S \subset N$. So, let us start by determining $v_\Gamma(S)$.

$$\begin{aligned}
v_\Gamma(S) &= \max\left\{\sum_{i\in S} m_i(d_i + R_i\boldsymbol{X}^N)\middle|\ (d_i + R_i\boldsymbol{X}^N)_{i\in N}\in \mathcal{Z}_\Gamma(N)\right\}\\
&= \sum_{j\in S_P}\tfrac{1}{\alpha_j}\log\left(\int_0^\infty e^{t/\alpha(S_I\cup\{j\})}dF_{\boldsymbol{X}_j}(t)\right)^{-1}\\
&\quad+\sum_{i\in S_I}\sum_{j\in S}\tfrac{1}{\alpha_i}\log\left(\int_0^\infty e^{t/\alpha(S_I)}dF_{\boldsymbol{X}_j}(t)\right)^{-1}\\
&= \sum_{j\in S_P}\tfrac{1}{\alpha_j}\log\left(\int_0^\infty e^{t/\alpha(S_I\cup\{j\})}dF_{\boldsymbol{X}_j}(t)\right)^{-1}\\
&\quad+\sum_{i\in S_I}\sum_{j\in S_P}\tfrac{1}{\alpha_i}\log\left(\int_0^\infty e^{t/\alpha(S_I)}dF_{\boldsymbol{X}_j}(t)\right)^{-1}\\
&= +\sum_{i\in S_I}\sum_{j\in S_I}\tfrac{1}{\alpha_i}\log\left(\int_0^\infty e^{t/\alpha(S_I)}dF_{\boldsymbol{X}_j}(t)\right)^{-1}\\
&= \sum_{j\in S_P}\sum_{i\in S_I}\cup\{j\}\tfrac{1}{\alpha_j}\log\left(\int_0^\infty e^{t/\alpha(S_I\cup\{j\})}dF_{\boldsymbol{X}_j}(t)\right)^{-1}\\
&\quad+\sum_{j\in S_I}\sum_{i\in S_I}\tfrac{1}{\alpha_i}\log\left(\int_0^\infty e^{t/\alpha(S_I)}dF_{\boldsymbol{X}_j}(t)\right)^{-1}\\
&= \sum_{j\in S_P}\alpha(S_I\cup\{j\})\log\left(\int_0^\infty e^{t/\alpha(S_I\cup\{j\})}dF_{\boldsymbol{X}_j}(t)\right)^{-1}\\
&\quad+\sum_{j\in S_I}\alpha(S_I)\log\left(\int_0^\infty e^{t/\alpha(S_I)}dF_{\boldsymbol{X}_j}(t)\right)^{-1}\\
&= \sum_{j\in S_P}\log\left(\int_0^\infty e^{t/\alpha(S_I\cup\{j\})}dF_{\boldsymbol{X}_j}(t)\right)^{-\alpha(S_I\cup\{j\})}\\
&\quad+\sum_{j\in S_I}\log\left(\int_0^\infty e^{t/\alpha(S_I)}dF_{\boldsymbol{X}_j}(t)\right)^{-\alpha(S_I)},
\end{aligned}\tag{6.6}$$

where the second equality follows from Theorem 6.3.

Next, note that for $i \in N_I$ we have that

$$\begin{aligned}
m_i(d_i^0 - R_i^*\boldsymbol{X}^N) &= d_i^0 + \sum_{j\in N_I\cup N_P} m_i(-r_{ij}^*\boldsymbol{X}_j)\\
&= -\sum_{j\in N_I\cup N_P} m_i(-r_{ij}^*\boldsymbol{X}_j) + \sum_{j\in N_I} m_j(-r_{ji}^*\boldsymbol{X}_i)\\
&\quad+\sum_{j\in N_I\cup N_P} m_i(-r_{ij}^*\boldsymbol{X}_j)\\
&= \sum_{j\in N_I} m_j(-r_{ji}^*\boldsymbol{X}_i)\\
&= \sum_{j\in N_I}\tfrac{1}{\alpha_j}\log\left(\int_0^\infty e^{t/\alpha(N_I)}dF_{\boldsymbol{X}_j}\right)^{-1}\\
&= \log\left(\int_0^\infty e^{t/\alpha(N_I)}dF_{\boldsymbol{X}_j}\right)^{-\alpha(N_I)}
\end{aligned}$$

and for $i \in N_P$ that

$$\begin{aligned} m_i(d_i^0 + R_i^* \boldsymbol{X}^N) &= d_i^0 + m_i(-r_{ii}^* \boldsymbol{X}_i) \\ &= \sum_{j \in N_I} m_j(-r_{ji}^* \boldsymbol{X}_i) + m_i(-r_{ii}^* \boldsymbol{X}_i) \\ &= \sum_{j \in N_I \cup \{i\}} m_j(-r_{ji}^* \boldsymbol{X}_i) \\ &= \sum_{j \in N_I \cup \{i\}} \tfrac{1}{\alpha_j} \log \left(\int_0^\infty e^{t/\alpha(N_I \cup \{i\})} dF_{\boldsymbol{X}_j} \right)^{-1} \\ &= \log \left(\int_0^\infty e^{t/\alpha(N_I \cup \{i\})} dF_{\boldsymbol{X}_j} \right)^{-\alpha(N_I \cup \{i\})}, \end{aligned}$$

Then for $S \subset N_I \cup N_P$ it holds that

$$\begin{aligned} \sum_{i \in S} m_i(d_i^0 + R_i^* \boldsymbol{X}^N) &= \sum_{i \in S_I} \log \left(\int_0^\infty e^{t/\alpha(N_I)} dF_{\boldsymbol{X}_j} \right)^{-\alpha(N_I)} \\ &\quad + \sum_{i \in S_P} \log \left(\int_0^\infty e^{t/\alpha(N_I \cup \{i\})} dF_{\boldsymbol{X}_j} \right)^{-\alpha(N_I \cup \{i\})} \\ &\geq \sum_{i \in S_I} \log \left(\int_0^\infty e^{t/\alpha(S_I)} dF_{\boldsymbol{X}_j} \right)^{-\alpha(S_I)} \\ &\quad + \sum_{i \in S_P} \log \left(\int_0^\infty e^{t/\alpha(S_I \cup \{i\})} dF_{\boldsymbol{X}_j} \right)^{-\alpha(S_I \cup \{i\})} \\ &= v_\Gamma(S), \end{aligned}$$

where the inequality follows from Lemma 6.7 with $c = 1$ and $x = \frac{1}{\alpha(S)}$. □

Example 6.5 In this example all monetary amounts are stated in thousands of dollars. Consider the following situation in automobile insurance with two insurance companies and three individual persons. So, $N_I = \{1, 2\}$ and $N_P = \{3, 4, 5\}$. The utility function of each agent can be described by $U_i(t) = e^{-\alpha_i t}$ with $\alpha_1 = 0.1$, $\alpha_2 = 0.167$, $\alpha_3 = 0.333$, $\alpha_4 = 0.125$ and $\alpha_5 = 0.2$, respectively. For evaluating random payoffs $\boldsymbol{X} \in L^1(\mathbb{R})$, we focus on the corresponding certainty equivalent $m_i(\boldsymbol{X}) = U_i^{-1}(E(U_i(\boldsymbol{X})))$, $i \in \{1, 2, 3, 4, 5\}$.

Each insurance company bears the risk of all the cars contained in its insurance portfolio. A car can be either one of two types. The first type corresponds to an average saloon car with a retail price of \$20, and generates relatively low losses. The second type corresponds to an exclusive sportscar with a retail price of \$200, and generates relatively high losses. More precisely, the monetary loss generated by a car is uniformly distributed between zero and its retail price. Thus the expected loss of a type 1 car and a type 2 car equal \$ 10 and \$ 100, respectively.

The insurance portfolio of insurer 1 consists of 900 cars of type 1 and 25 cars of type 2. For insurer 2 the portfolio consists of 400 cars of type 1 and 70 cars of type 2. The expected loss for insurer 1 then equals $900 \cdot 10 + 25 \cdot 100 = \$\,11500$. The expected losses for insurer 2 equals \$ 11000. The individuals 3 and 4 each possess one car. Individual 3's car is of type 1 and individual 4's car is of type 2. Individual 5 possesses both cars. The expected losses are \$ 10, \$ 100, and \$ 110, respectively.

Next, let $\boldsymbol{X}_i$ denote the loss of agent i. If all agents cooperate, the Pareto optimal risk allocation matrix of the total random loss $\boldsymbol{X}_1+\boldsymbol{X}_2+\boldsymbol{X}_3+\boldsymbol{X}_4+\boldsymbol{X}_5$ equals

$$R^* = \begin{bmatrix} \frac{5}{8} & \frac{5}{8} & \frac{10}{19} & \frac{5}{12} & \frac{10}{21} \\ \frac{3}{8} & \frac{3}{8} & \frac{6}{19} & \frac{3}{12} & \frac{6}{21} \\ 0 & 0 & \frac{3}{19} & 0 & 0 \\ 0 & 0 & 0 & \frac{4}{12} & 0 \\ 0 & 0 & 0 & 0 & \frac{5}{21} \end{bmatrix},$$

Then the certainty equivalent of a Pareto optimal allocation $(d_i + R_i^* \boldsymbol{X}^N)_{i\in N} \in \mathcal{Z}_\Gamma(N)$ equals

$$\begin{aligned} m_1(d_1 + R_1^* \boldsymbol{X}^N) &= d_1 - 18582, \\ m_2(d_2 + R_2^* \boldsymbol{X}^N) &= d_2 - 11149, \\ m_3(d_3 + R_3^* \boldsymbol{X}^N) &= d_3 - 2, \\ m_4(d_4 + R_4^* \boldsymbol{X}^N) &= d_4 - 50, \\ m_5(d_5 + R_5^* \boldsymbol{X}^N) &= d_5 - 39. \end{aligned}$$

The insurance premiums that the agents have to pay, are calculated according to the zero utility principle. This means that the aggregate premium which insurer 1 receives, equals

$$\begin{aligned} d_1^0 &= \pi_1\left(\tfrac{10}{16}\boldsymbol{X}_2\right) + \pi_1\left(\tfrac{10}{19}\boldsymbol{X}_3\right) + \pi_1\left(\tfrac{10}{24}\boldsymbol{X}_4\right) + \pi_1\left(\tfrac{10}{21}\boldsymbol{X}_5\right) - \pi_2\left(\tfrac{6}{16}\boldsymbol{X}_1\right) \\ &= -m_1\left(-\tfrac{10}{16}\boldsymbol{X}_2\right) - m_1\left(-\tfrac{10}{19}\boldsymbol{X}_3\right) - m_1\left(-\tfrac{10}{24}\boldsymbol{X}_4\right) \\ &\quad -m_1\left(-\tfrac{10}{21}\boldsymbol{X}_5\right) + m_2\left(-\tfrac{6}{16}\boldsymbol{X}_1\right) \\ &\approx 9739 + 6 + 62 + 78 - 5597 = 4288. \end{aligned}$$

Similarly, we get for insurer 2 and individuals 3, 4 and 5

$$
\begin{aligned}
d_2^0 &= 5597 + 3 + 37 + 47 - 9739 \;=\; -4055, \\
d_3^0 &= -3 - 6 \;=\; -9, \\
d_4^0 &= -37 - 62 \;=\; -99, \\
d_5^0 &= -47 - 78 \;=\; -125.
\end{aligned}
$$

With $d^0 = (4288, -4055, -9, -99, -125)$ being the aggregate insurance premiums we obtain that $(m_i(d_i^0 + R_i^* \boldsymbol{X}^N))_{i \in N} = (-14294, -15204, -11, -149, -164)$. It is a straightforward exercise to check that $(-14294, -15204, -11, -149, -164)$ is a core allocation of the corresponding TU-game (N, v_Γ) presented in Table 6.1.

S	$v_\Gamma(S)$	S	$v_\Gamma(S)$	S	$v_\Gamma(S)$
$\{1\}$	-14704	$\{2,5\}$	-17730	$\{2,3,4\}$	-17725
$\{2\}$	-17551	$\{3,4\}$	-189	$\{2,3,5\}$	-17742
$\{3\}$	-14	$\{3,5\}$	-209	$\{2,4,5\}$	-17893
$\{4\}$	-174	$\{4,5\}$	-369	$\{3,4,5\}$	-383
$\{5\}$	-195	$\{1,2,3\}$	-29509	$\{1,2,3,4\}$	-29658
$\{1,2\}$	-29498	$\{1,2,4\}$	-29647	$\{1,2,3,5\}$	-29672
$\{1,3\}$	-14715	$\{1,2,5\}$	-29661	$\{1,2,4,5\}$	-29810
$\{1,4\}$	-14861	$\{1,3,4\}$	-14872	$\{1,3,4,5\}$	-15044
$\{1,5\}$	-14876	$\{1,3,5\}$	-14888	$\{2,3,4,5\}$	-17905
$\{2,3\}$	-17562	$\{1,4,5\}$	-15053	$\{1,2,3,4,5\}$	-29822
$\{2,4\}$	-17713				

Table 6.1

Now, let us take a closer look at the changes in insurer 1's utility when the allocation $(d_i^0 + R_i^* \boldsymbol{X}^N)_{i \in N}$ is realized. In the initial situation insurer 1 bears the risk $\boldsymbol{X}_1$ of his own insurance portfolio. The certainty equivalent of $\boldsymbol{X}_1$ equals

$$
\begin{aligned}
m_1(\boldsymbol{X}_1) &= 900 \cdot -10 \log\left(0.5(e^2 - 1)\right) + 25 \cdot -10 \log\left(0.05(e^{20} - 1)\right) \\
&\approx -14704.
\end{aligned}
$$

To allocate the total risk in a Pareto optimal way, insurer 1 bears the fraction $r_{12}^* = \frac{10}{16}$ of the risk $\boldsymbol{X}_2$ of insurer 2. For this risk he receives a premium $\pi_1(\frac{10}{16}\boldsymbol{X}_2)$ as determined by the zero utility principle. From the definition of the zero utility calculation principle and the independence of $\boldsymbol{X}_1$ and $\boldsymbol{X}_2$, it follows that $m_1(-\boldsymbol{X}_1 - \frac{10}{16}\boldsymbol{X}_2 + \pi_1(\frac{10}{16}\boldsymbol{X}_2)) \approx -14704$. So insurer 1's welfare does not change when he insures a part of the risk of insurer 2. A similar argument holds when he insures a part of the risks of the other agents. Hence

$$
\begin{aligned}
&m_1(-\boldsymbol{X}_1 - \tfrac{10}{16}\boldsymbol{X}_2 + \pi_1(\tfrac{10}{16}\boldsymbol{X}_2) - \tfrac{10}{19}\boldsymbol{X}_3 + \pi_1(\tfrac{10}{19}\boldsymbol{X}_3) \\
&\quad - \tfrac{10}{24}\boldsymbol{X}_4 + \pi_1(\tfrac{10}{24}\boldsymbol{X}_4) - \tfrac{10}{21}\boldsymbol{X}_5 + \pi_1(\tfrac{10}{21}\boldsymbol{X}_5)) \;\approx\; -14704
\end{aligned}
$$

The increase in insurer 1's welfare arises only from the risk $\frac{6}{16}\boldsymbol{X}_1$ he transfers to insurers 2:

$$\begin{aligned} & m_1(-\tfrac{10}{16}\boldsymbol{X} - \pi_2(\tfrac{6}{16}\boldsymbol{X}_1) - \tfrac{10}{16}\boldsymbol{X}_2 + \pi_1(\tfrac{10}{16}\boldsymbol{X}_2) - \tfrac{10}{19}\boldsymbol{X}_3 + p_1(\tfrac{10}{19}\boldsymbol{X}_3) \\ & \quad -\tfrac{10}{24}\boldsymbol{X}_4 + \pi_1(\tfrac{10}{24}\boldsymbol{X}_4) - \tfrac{10}{21}\boldsymbol{X}_5 + \pi_1(\tfrac{10}{21}\boldsymbol{X}_5)) \\ = \; & m_1(d_1^0 + R_1^*\boldsymbol{X}^N) \; \approx \; -14294 \; > \; -14704. \end{aligned}$$

The phenomenon described above is subsistent in the definition of the zero utility principle. This means that the welfare of an insurer always remains the same when he bears the risk of someone else in exchange for the zero utility principle based premium. An increase in welfare only arises when he transfers (a part of) his own risk to someone else.

6.2 Subadditivity for Collective Insurances

In the insurance games defined in the previous section individual persons are not allowed to cooperate; they cannot redistribute the risk amongst themselves. Looking at the individuals' behavior in everyday life, this is a justified assumption. People who want to insure themselves against certain risks do so by contacting insurance companies, pension funds etc. We show, however, that when this restriction is abandoned then the mere fact that risk exchanges could take place between individuals implies that insurance companies have incentives to employ subadditive premiums. Whether or not such risk exchanges actually do take place is not important. As a consequence, collective insurances become cheaper for the individuals.

Let N_P be the set of individuals. A premium calculation principle π is called subadditive if for all subsets $S, T \subset N_P$ with $S \cap T = \emptyset$ it holds that $\pi(\boldsymbol{X}_S) + \pi(\boldsymbol{X}_T) \geq \pi(\boldsymbol{X}_S + \boldsymbol{X}_T)$. Here, $\boldsymbol{X}_S$ denotes the total loss of the coalition S. So, it is attractive for the individuals to take a collective insurance, since this reduces the total premium they have to pay.

Next, consider a game with agent set N_P only where the individuals are allowed to redistribute their risks. This situation can be described by an insurance game $\Gamma \in IG(N_P, \emptyset)$. So, the individuals N_P can now insure their losses among each other. Then we can associate with Γ the TU-game (N, v_Γ), with

$$v_\Gamma(S) = \max\left\{\sum_{i \in S} m_i(d_i - R_i\boldsymbol{X}^S) | \, (d_i - R_i\boldsymbol{X}^S)_{i \in S} \in \mathcal{Z}_\Gamma(S)\right\}$$

for all $S \subset N_P$. Note that this maximum is attained for Pareto optimal allocations $(d_i - R_i^*\boldsymbol{X}^S)_{i \in S} \in \mathcal{Z}_\Gamma(S)$ for coalition S. For this game, the value $v_\Gamma(S)$ can be interpreted as the maximum premium coalition S wants to pay for the insurance of the total risk $\boldsymbol{X}_S$. To see this, suppose that the coalition S can insure the loss $\boldsymbol{X}_S$ for a premium $\pi(\boldsymbol{X}_S)$ that exceeds the valuation of the risk $\boldsymbol{X}_S$, that is, $-\pi(\boldsymbol{X}_S) < v_\Gamma(S)$. Then for each allocation $y \in \mathbb{R}^S$ of the premium $-\pi(\boldsymbol{X}_S)$ there exists an allocation $(\tilde{d}_i - R_i^* \boldsymbol{X}^S) \in \mathcal{Z}_\Gamma(S)$ such that $E(U_i(\tilde{d}_i - R_i^* \boldsymbol{X}^S)) > U_i(y_i)$ for all $i \in S$. Indeed, let $(d_i - R_i^* \boldsymbol{X}^S)_{i \in S} \in \mathcal{Z}_\Gamma(S)$ be such that $\sum_{i \in S} m_i(d_i - R_i^* \boldsymbol{X}^S) = v_\Gamma(S)$. Define

$$\tilde{d}_i = d_i - m_i(d_i - R_i^* \boldsymbol{X}^S) + y_i + \tfrac{1}{\#S}\left(v_\Gamma(S) + \pi(\boldsymbol{X}_S)\right),$$

for all $i \in S$. Since

$$\begin{aligned}\sum_{i \in S} \tilde{d}_i &= \sum_{i \in S} d_i - \sum_{i \in S} m_i(d_i - R_i \boldsymbol{X}^S) + \sum_{i \in S} y_i + v_\Gamma(S) + \pi(\boldsymbol{X}_S) \\ &= \textstyle\sum_{i \in S} d_i \leq 0\end{aligned}$$

it follows that $(\tilde{d}_i - R_i \boldsymbol{X}^S)_{i \in S} \in \mathcal{Z}_\Gamma(S)$. Then by the linearity of m_i in $\tilde{d}_i$ (cf. expression (6.2)) we have for all $i \in S$ that

$$m_i(\tilde{d}_i - R_i^* \boldsymbol{X}^S) = y_i + \tfrac{1}{\#S}\left(v_\Gamma(S) + \pi(\boldsymbol{X}_S)\right) > y_i.$$

Hence, the members of S prefer the allocation $(\tilde{d}_i - R_i^* \boldsymbol{X}^S)_{i \in S}$ of $\boldsymbol{X}_S$ to an insurance of $\boldsymbol{X}_S$ and paying the premium $\pi(\boldsymbol{X}_S)$. Consequently, they will not pay more for the insurance of the risk $\boldsymbol{X}_S$ than the amount $-v_\Gamma(S)$. Now, it is a straightforward exercise to show that this maximum premium $-v_\Gamma(S)$ satisfies subadditivity, i.e. $-v_\Gamma(S) - v_\Gamma(T) \geq -v_\Gamma(S \cup T)$. For totally balancedness of insurance games implies superadditivity, i.e. $v_\Gamma(S) + v_\Gamma(T) \leq v_\Gamma(S \cup T)$ for all disjoint $S, T \subset N$.

6.3 Remarks

In this chapter (re)insurance problems are modelled as cooperative games with stochastic payoffs. In fact, we defined a game that dealt with both the insurance and the reinsurance problem simultaneously. We showed that there is only one allocation risk exchange matrix yielding a Pareto optimal distribution of the losses and that a core allocation results when insurance premiums are calculated according to the zero utility principle.

Recall that insurers do not benefit from insuring the risks of the individuals when utilizing the additive zero utility principle; this premium calculation principle yields the lowest premium for which insurers still want to exchange risks with the individuals (see Example 6.5). So, from a social point of view, it might be best to adopt a middle course and look for premiums where both insurers and individuals benefit from the insurance transaction. Interesting questions then remaining are: are these premiums additive or subadditive and do they yield core allocations?

An issue only briefly mentioned in this paper concerns the insurers' behavior. What if an insurer is risk neutral or risk loving instead of risk averse? Thus, there is at least one insurer whose utility function is linear or of the form $u_i(t) = \beta_i e^{-\alpha_i t}$ ($t \in \mathbb{R}$) with $\beta_i > 0$, $\alpha_i < 0$. Although the proofs are not provided here, most of the results presented in this paper still hold for these situations. This means that the corresponding games have nonempty cores and that the zero utility principle still yields a core allocation. The result that does change is the Pareto optimal allocation of the risk. The allocations that are Pareto optimal when all insurers are risk averse are not Pareto optimal anymore when one or more insurers happen to be risk loving. In fact, they are the worst possible allocations of the risk one can think of. In that case, allocating all the risk to the most risk loving insurer is Pareto optimal. This would actually mean that only one insurance company is needed, since other insurance companies will ultimately reinsure their complete portfolios at this most risk loving insurer.

Finally, it should be mentioned that the results presented in this chapter still go through if we replace the assumption that the risks $(\boldsymbol{X}_i)_{i \in N_P \cup N_i}$ are mutually independent by the assumption that the covariance matrix of $(\boldsymbol{X}_i)_{i \in N_P \cup N_i}$ is negative definite.

6.4 Appendix: Proofs

Lemma 6.6 Let $c \in \mathbb{R} \backslash \{0\}$ and let F be a probability distribution function corresponding to a non-degenerate random variable. Then the function h_c defined by $h_c(x) = \log \left(\int_0^\infty e^{xct} dF(t) \right)^{-1}$ for $x \geq 0$, is strictly concave in x.

PROOF: Since

$$\frac{dh_c}{dx} = -c \frac{\int_0^\infty t e^{xct} dF(t)}{\int_0^\infty e^{xct} dF(t)}$$

we obtain that

$$\begin{aligned} \frac{dh_c^2}{dx^2} &= -c^2 \frac{\int_0^\infty t^2 e^{xct} dF(t) \int_0^\infty e^{xct} dF(t) - \left(\int_0^\infty t e^{xct} dF(t) \right)^2}{\left(\int_0^\infty e^{xct} dF(t) \right)^2} \\ &= -c^2 \int_0^\infty \left(t - \frac{\int_0^\infty \tau e^{xc\tau} dF(\tau)}{\int_0^\infty e^{xc\tau} dF(\tau)} \right)^2 \frac{e^{xct}}{\int_0^\infty e^{xc\tau} dF(\tau)} dF(t) \\ &\leq 0. \end{aligned}$$

Hence, h_c is concave. The lemma then follows from the observation that the inequality is binding if and only if F corresponds to a degenerate random variable.□

Lemma 6.7 Let $c \in \mathbb{R}\backslash\{0\}$. Then the function $h_c(x) = \left(\int_0^\infty e^{xct} dF(t)\right)^{\frac{1}{x}}$, $x > 0$, is increasing in x.

PROOF: Note that

$$\begin{aligned}
\frac{dh_c}{dx} &= \frac{d}{dx} e^{\frac{1}{x}\log\left(\int_0^\infty e^{xct} dF(t)\right)} \\
&= h_c(x)\left(\frac{-1}{x^2}\log\left(\int_0^\infty e^{xct} dF(t)\right) + \frac{c}{x}\frac{\left(\int_0^\infty te^{xct} dF(t)\right)}{\left(\int_0^\infty e^{xct} dF(t)\right)}\right) \\
&= -\frac{h_c(x)}{x^2}\left(\log\left(\int_0^\infty e^{xct} dF(t)\right) - xc\frac{\left(\int_0^\infty te^{xct} dF(t)\right)}{\left(\int_0^\infty e^{xct} dF(t)\right)}\right).
\end{aligned}$$

Since $x > 0$ it is sufficient to show that

$$\log\left(\int_0^\infty e^{xct} dF(t)\right) - xc\frac{\left(\int_0^\infty te^{xct} dF(t)\right)}{\left(\int_0^\infty e^{xct} dF(t)\right)} \leq 0. \tag{6.7}$$

First, note that

$$\lim_{x \downarrow 0} \log\left(\int_0^\infty e^{xct} dF(t)\right) - xc\frac{\left(\int_0^\infty te^{xct} dF(t)\right)}{\left(\int_0^\infty e^{xct} dF(t)\right)} = 0.$$

Second, differentiating expression (6.7) to x yields

$$\begin{aligned}
& c\frac{\int_0^\infty te^{xct} dF(t)}{\int_0^\infty e^{xct} dF(t)} - c\frac{\int_0^\infty te^{xct} dF(t)}{\int_0^\infty e^{xct} dF(t)} - xc\frac{d}{dx}\left(\frac{\int_0^\infty te^{xct} dF(t)}{\int_0^\infty e^{xct} dF(t)}\right) \\
&= -xc\frac{d}{dx}\frac{\int_0^\infty te^{xct} dF(t)}{\int_0^\infty e^{xct} dF(t)} \\
&= -xc^2\frac{\int_0^\infty t^2 e^{xct} dF(t) \int_0^\infty e^{xct} dF(t) - \left(\int_0^\infty te^{xct} dF(t)\right)^2}{\left(\int_0^\infty e^{xct} dF(t)\right)^2} \\
&= -xc^2 \int_0^\infty \left(t - \frac{\int_0^\infty \tau e^{xc\tau} dF(\tau)}{\int_0^\infty e^{xc\tau} dF(\tau)}\right)^2 \frac{e^{xct}}{\int_0^\infty e^{xc\tau} dF(\tau)} dF(t) \\
&\leq 0.
\end{aligned}$$

Thus, since (6.7) is a decreasing function in x taking the value zero in $x = 0$, it follows that (6.7) is negative for all $x > 0$. □

7 Price Uncertainty in Linear Production Situations

Production processes are classic examples of situations where several parties recognize the benefits of cooperation. The owners of the production factors labor, resources, capital, and technology join forces to deliver a product or a service that is valued more by consumers than their separate inputs. The value that is added by production is the benefit of cooperation and has to be divided between the parties that are involved in the production process.

Owen (1975) was the first to analyze this problem from a game theoretical point of view. In his model, agents have their own individual bundle of resources which they can use as inputs for a publicly available linear production technology. The resulting output can be sold on the market for given prices. In this situation, agents can individually use their resources to maximize the proceeds from production, but they can probably do better if they cooperate with each other and combine their resources. For instance, an agent that lacks certain inputs for production, prefers cooperating with agents that possess the required inputs, so that production can take place. Owen (1975) shows that the corresponding linear production games are totally balanced. In particular, he shows that a core-allocation arises if each agent receives the marginal value of his resources. Here, the marginal value of a resource is the marginal revenue generated by one extra unit of this resource.

The analysis of Owen (1975) as well as most subsequent studies on this subject, confine themselves to a deterministic setting. Real production, however, typically features uncertainty. Due to irregularities in the production process, the quality of produced output may not always be up to standard so that production losses may occur. If this is the case, there is uncertainty about the output. Similarly, when production takes a considerable amount of time, there is uncertainty about the price at the moment that the production decision is made.

Sandsmark (1999) examines the cooperative behavior in a two stage production model with uncertainty about the output. The first stage production plan is determined under uncertainty while the second stage production plan may be contingent on the realized state of nature. It is shown that the resulting cooperative game, in which coalitions maximize a certain revenuefunction, has a nonempty core.

Furthermore, a specific core-allocation is provided. Similar to chance-constrained games, this model does not explicitly take into account the individual preferences of the agents. The revenuefunction, however, may be interpreted as the sum of individual expected utilities.

In contrast to Sandsmark (1999), this chapter analyzes linear production situations with price uncertainty. When agents decide upon their production plan, prices are still unknown. As a result, price volatility may play a role in deciding how agents use their resources. For instance, they may use their resources to produce different outputs so as to reduce the variability in total revenues. Particularly interesting in this regard is, that linear production games with price uncertainty can be used to describe investment problems, where agents can apply their (individual) capital to invest in financial assets whose future value is uncertain at the time of investment. (see also Example 3.3). We discuss this application more detailed in a separate section.

7.1 Stochastic Linear Production Games

In line with the notation of Example 2.2 on linear production situations, let $N \subset \mathbb{N}$ denote the set of agents, $R \subset \mathbb{N}$ the set of resources, and $M \subset \mathbb{N}$ the set of consumption goods. By assumption, each agent $i \in N$ is a risk averse expected utility maximizer with utility function $U_i(t) = \beta_i e^{-\alpha_i t}$, where $\beta_i < 0$ and $\alpha_i > 0$. Let $b^i \in \mathbb{R}^R_+$ denote agent i's endowment of resources, and let $A^i \in \mathbb{R}^{R\times M}$ denote the linear production technology of agent i. This means that for the production of an output bundle $c \in \mathbb{R}^R_+$ the inputs $A^i c \in \mathbb{R}^M_+$ are needed. Note that this generalizes Owen (1975) and Example 2.2, since in those cases each agent has access to the same technology $A \in \mathbb{R}^{R\times M}$. The resources can be used to produce consumption goods. The prices at which these goods can be sold, are denoted by random variables $\boldsymbol{P}_j \in L^1(\mathbb{R}_+)$, $j \in M$, and are assumed to be mutually independent.

When coalition S forms, the members of S pool their resources and production technologies, so that they possess the resources $\sum_{i\in S} b^i$ and have access to the technologies A^i, $i \in S$. Let $c^i \in \mathbb{R}^R_+$ denote the consumption bundle that coalition S plans to produce with the production technology A^i of agent $i \in S$. Then a production plan $c \in \mathbb{R}^R_+$ is feasible for coalition S if it possesses the necessary resources, that is there exist separate production plans $(c^i)_{i\in S}$ such that $\sum_{i\in S} c^i = c$ and $\sum_{i\in S} A^i c^i \leq \sum_{i\in S} b^i$. The set of feasible production plans for coalition $S \subset N$ thus equals

$$C(S) = \left\{ \sum_{i \in S} c^i \middle| \forall_{i \in S} : c^i \in \mathbb{R}^R_+, \sum_{i \in S} A^i c^i \leq \sum_{i \in S} b^i \right\}. \tag{7.1}$$

Each feasible production plan $c \in C(S)$ yields stochastic revenues $\sum_{j \in M} \boldsymbol{P}_j c_j$. Now, we can describe a linear production situation with price uncertainty by the following *stochastic linear production game* $\Gamma = (N, \mathcal{V}, \{\succsim\}_{i \in N})$, with

$$\mathcal{V}(S) = \left\{ \sum_{j \in M} \boldsymbol{P}_j c_j \middle| c \in C(S) \right\} \tag{7.2}$$

for all $S \subset N$ and $\succsim_i$ the preferences induced by U_i, for each $i \in N$. The set of all stochastic linear production games is denoted by $SLP(N)$.

As was the case for insurance games, exponential utility functions imply that stochastic linear production games belong to the class $MG(N)$. Hence, for our analysis we may focus on certainty equivalents. Therefore, let $\Gamma \in SLP(N)$ be a stochastic linear production game. Then, given a feasible production plan $c \in C(S)$, the certainty equivalent of an allocation $\left(d_i + r_i \sum_{j \in M} \boldsymbol{P}_j c_j\right)_{i \in S}$ is for each agent $i \in S$ equal to

$$\begin{aligned} m_i\left(d_i + r_i \sum_{j \in M} \boldsymbol{P}_j c_j\right) &= U_i^{-1}\left(E(U_i(d_i + r_i \sum_{j \in M} \boldsymbol{P}_j c_j))\right) \\ &= d_i + \sum_{j \in M} U_i^{-1}\left(E\left(U_i\left(r_i \boldsymbol{P}_j c_j\right)\right)\right) \\ &= d_i + \sum_{j \in M} \tfrac{1}{\alpha_i} \sum_{j \in M} \log\left(\int_0^\infty e^{-\alpha_i r_i c_j t} dF_{\boldsymbol{P}_j}(t)\right)^{-1}, \end{aligned}$$

where the second equality follows from the independence of $(\boldsymbol{P}_j)_{j \in M}$. The corresponding TU-game (N, v_Γ) is then given by

$$\begin{aligned} v_\Gamma(S) &= \max \sum_{i \in S} \tfrac{1}{\alpha_i} \sum_{j \in M} \log\left(\int_0^\infty e^{-\alpha_i r_i c_j t} dF_{\boldsymbol{P}_j}(t)\right)^{-1} \\ \text{s.t.:} \quad & \textstyle\sum_{i \in S} r_i = 1, \\ & r_i \geq 0, \quad \text{for all } i \in S, \\ & c \in C(S), \end{aligned} \tag{7.3}$$

for all $S \subset N$.

We cannot explicitly determine an optimal production plan $c \in C(S)$. What we can do though, is determine the Pareto optimal allocation r of the random revenue $\sum_{j \in M} \boldsymbol{P}_j c_j$, while taking the production plan $c \in C(S)$ as given.

Proposition 7.1 Let $\Gamma \in SLP(N)$ and let $c \in C(S)$ be a feasible production plan for coalition $S \subset N$. An allocation $\left(d_i + r_i \sum_{j \in M} \boldsymbol{P}_j c_j\right)_{i \in S}$ is a Pareto optimal allocation of the revenue $\sum_{j \in M} \boldsymbol{P}_j c_j$ if and only if

$$r_i^* = \frac{\frac{1}{\alpha_i}}{\sum_{j\in S}\frac{1}{\alpha_j}} \tag{7.4}$$

for all $i \in S$.

PROOF: By Proposition 3.13, it suffices to show that r^* is the unique solution to

$$\begin{aligned} \max \sum_{i\in S} \tfrac{1}{\alpha_i} \sum_{j\in M} \log \left(\int_0^\infty e^{-\alpha_i r_i c_j t} dF_{\boldsymbol{P}_j}(t) \right)^{-1} & \\ \text{s.t.:} \quad \textstyle\sum_{i\in S} r_i \; = \; 1, & \\ r_i \; \geq \; 0, \quad \text{for all} i \in S. & \end{aligned}$$

From Lemma 6.6 with $c = -\alpha_i c_j$ and $x = r_i$ it follows that the objective function is strictly concave, so that the optimal solution is unique. That r^* is indeed the unique solution follows from the fact that it satisfies the Karush-Kuhn-Tucker conditions (see page 94):

$$\begin{aligned} -\sum_{j\in M} c_j \frac{\int_0^\infty t e^{-\alpha_i r_i^* c_j t} dF_{\boldsymbol{P}_j}(t)}{\int_0^\infty e^{-\alpha_i r_i^* c_j t} dF_{\boldsymbol{P}_j}(t)} &= \lambda - \nu_i, \\ \nu_i r_i &= 0, \qquad \text{for all } i \in S, \end{aligned}$$

with $\nu_i = 0$ for all $i \in S$,

$$\lambda = -\sum_{j\in M} c_j \frac{\int_0^\infty t e^{-c_j t/\alpha(S)} dF_{\boldsymbol{P}_j}(t)}{\int_0^\infty e^{-c_j t/\alpha(S)} dF_{\boldsymbol{P}_j}(t)},$$

and $\alpha(S) = \sum_{i\in S} \frac{1}{\alpha_i}$. □

Note that more risk averse agents bear a larger part of the risk. Furthermore, note that the Pareto optimal allocation r^* is independent of the production plan $c \in C(S)$, so that r^* also yields a Pareto optimal allocation for the optimal production plan.

The following theorem shows that stochastic linear production games are totally balanced.

Theorem 7.2 Each stochastic linear production game $\Gamma \in SLP(N)$ is totally balanced.

PROOF: Let $\Gamma \in SLP(N)$. Since each subgame $\Gamma_{|S} \in SLP(S)$, it suffices to show that Γ is balanced. For this, we apply Theorem 2.9 and Theorem 4.1. First, we substitute (7.4) in expression (7.3) and derive that

$$v_\Gamma(S) = \max \left\{ \alpha(S) \sum_{j\in M} \log \left(\int_0^\infty e^{-c_j t/\alpha(S)} dF_{\boldsymbol{P}_j}(t) \right)^{-1} \middle| \, c \in C(S) \right\}. \tag{7.5}$$

Second, let λ be a balanced map and denote the optimal production plan for coalition S by $c^S \in C(S)$. In particular, let $(c^{S,i})_{i\in S}$ be such that $\sum_{i\in S} c^{S,i} = c^S$ and $\sum_{i\in S} A^i c^{S,i} \leq \sum_{i\in S} b^i$. So, $c^{S,i}$ is that part of the production plan that is produced with the technology A^i. Since

$$\begin{aligned}
\sum_{i\in S} A^i \sum_{S\subset N} \lambda(S)c^{S,i} &= \sum_{S\subset N} \lambda(S) \sum_{i\in S} A^i c^{S,i} \\
&\leq \sum_{S\subset N} \lambda(S) \sum_{i\in S} b^i \\
&= \sum_{i\in N} \sum_{S\subset N: i\in S} \lambda(S) b^i \\
&= \sum_{i\in N} b^i,
\end{aligned}$$

it holds that $\sum_{S\subset N} \lambda(S)c^S \in C(N)$. Next, we derive that

$$\begin{aligned}
\sum_{S\subset N} \tfrac{\lambda(S)}{\alpha(N)} v_\Gamma(S) &= \sum_{S\subset N} \tfrac{\lambda(S)\alpha(S)}{\alpha(N)} \sum_{j\in M} \log\left(\int_0^\infty e^{-c_j^S t/\alpha(S)} dF_{\mathbf{P}_j}\right)^{-1} \\
&\leq \sum_{j\in M} \log\left(\int_0^\infty e^{-\sum_{S\subset N} \lambda(S)\alpha(S)/\alpha(N)\cdot c_j^S t/\alpha(S)} dF_{\mathbf{P}_j}(t)\right)^{-1} \\
&= \sum_{j\in M} \log\left(\int_0^\infty e^{-\sum_{S\subset N} \lambda(S)c_j^S t/\alpha(N)} dF_{\mathbf{P}_j}(t)\right)^{-1}, \qquad (7.6)
\end{aligned}$$

where the inequality follows from $\sum_{S\subset N} \frac{\lambda(S)\alpha(S)}{\alpha(N)} = 1$ and Lemma 6.6, which states that the objective function is concave. Balancedness then follows from

$$\begin{aligned}
\sum_{S\subset N} \lambda(S) v_\Gamma(S) &\leq \alpha(N) \sum_{j\in M}^{m} \log\left(\int_0^\infty e^{-\sum_{S\subset N} \lambda(S)c_j^S t/\alpha(N) dF_{\mathbf{P}_j}(t)}\right)^{-1} \\
&\leq \alpha(N) \sum_{j\in M} \log\left(\int_0^\infty e^{-c_j^N t/\alpha(N)} dF_{\mathbf{P}_j}(t)\right)^{-1} \\
&= v_\Gamma(N),
\end{aligned}$$

where the first and second inequality follow from (7.6) and $\sum_{S\subset N} \lambda(S)c^S \in C(N)$, respectively. □

With deterministic prices, a core-allocation can be found by appropriately valuating the different resources, and giving each agent the total value of his own resource bundle. To see how this process works, recall that a deterministic linear production game (N, v) is given by

$$v(S) = \max\left\{ p^\top c \,\middle|\, Ac \leq \sum_{i\in S} b^i \right\} \qquad (7.7)$$

for all $S \subset N$ (see Example 2.6). The dual optimization problem of (7.7) for coalition N is given by

$$\min\left\{x^\top \sum_{i\in N} b^i \,\middle|\, x^\top A \geq p\right\}.$$

Let $\xi \in \mathbb{R}_+^R$ be the optimal solution of the dual problem. Then ξ_k represents the marginal value of resource k, that is one extra unit of resource k raises the value $v(N)$ with ξ_k. One can interpret ξ_k as the value of owning one unit of resource k, and $\sum_{k\in R}\xi_k b_k$ as the total value of owning the resource bundle $b \in \mathbb{R}_+^R$. Owen (1975) shows that a core-allocation arises for the linear production game defined in (7.7), if each agent $i \in N$ receives the total value $\sum_{k\in R}\xi_k b_k^i$ of his individual resource bundle b^i.

This naturally raises the question if we can construct core-allocations in a similar way for stochastic linear production games? The answer is no. In general, there exists no $\xi \in \mathbb{R}_+^R$ such that $\left(\xi^\top b^i\right)_{i\in N}$ is a core-allocation for the stochastic linear production game (N, v_Γ) as defined in (7.3) and (7.5).

Example 7.3 Consider the following two person stochastic linear production game $\Gamma \in SLP(N)$, with $\alpha_1 = 0.1$, $\alpha_2 = 10$, $A = [1]$, $b^1 = 0.1$, $b^2 = 1$, and $\boldsymbol{P}_1$ exponentially distributed with parameter 0.5. Note that the production technology uses one unit of input to produce one unit of output. The corresponding TU-game (N, v_Γ) equals $v_\Gamma(\{1\}) = 0.198$, $v_\Gamma(\{2\}) = 1.019$, and $v_\Gamma(\{1,2\}) = 1.990$. Now, let $\xi \geq 0$ denote the value of the single resource. If $(\xi b^1, \xi b^2)$ belongs to the core of (N, v_Γ), it holds that $\xi b^1 = 0.1\xi \geq 0.198 = v_\Gamma(\{1\})$, $\xi b^2 = \xi \geq 1.019 = v_\Gamma(\{2\})$, and $\xi b^1 + \xi b^2 = 1.1\xi = 1.990 = v_\Gamma(\{1,2\})$. Since the first inequality implies that $\xi \geq 1.98$, we obtain the contradiction that $1.1\xi \geq 2.178 > 1.990 = v_\Gamma(\{1,2\})$. Hence, we cannot obtain a core-allocation by appropriately valuating the resource.

Theorem 7.4 Let $\Gamma \in SLP(N)$. If the resource bundles $(b^i)_{i\in N}$ are linearly independent, then there exists a vector $\xi \in \mathbb{R}^R$ such that $(\xi^\top b^i)_{i\in N} \in C(v_\Gamma)$.

PROOF: Take $\Gamma \in SLP(N)$ and let $\xi \in \mathbb{R}^R$. Then $(\xi^\top b^i)_{i\in N} \in C(v_\Gamma)$ if

$$\begin{aligned} -\xi^\top \sum_{i\in S} b^i &\leq -v_\Gamma(S) \quad \text{for all } S \subset N,\\ \xi^\top \sum_{i\in N} b^i &\geq v_\Gamma(N). \end{aligned} \tag{7.8}$$

By using a variant of Farka's Lemma[1], such ξ exists if and only if there exist no $\mu \geq 0$, $\lambda(S) \geq 0$ for all $S \subset N$, such that

$$\sum_{S \subset N} -\lambda(S) \sum_{i \in S} b^i + \mu \sum_{i \in N} b^i = 0,$$
$$\sum_{S \subset N} -\lambda(S) v_\Gamma(S) + \mu v_\Gamma(N) < 0.$$

Rearranging terms yields that there exist no $\mu \geq 0$, $\lambda(S) \geq 0$ for all $S \subset N$, satisfying

$$\sum_{i \in N} b^i \sum_{S \subset N : i \in S} \lambda(S) = \mu \sum_{i \in N} b^i,$$
$$\sum_{S \subset N} \lambda(S) v_\Gamma(S) > \mu v_\Gamma(N).$$

Since $\lambda(S) \geq 0$, $v_\Gamma(S) \geq 0$ for all $S \subset N$, and $b^i \geq 0$ for all $i \in N$, it follows that $\mu > 0$. Hence, without loss of generality we may assume that $\mu = 1$, that is there exist no $\lambda(S) \geq 0$ for all $S \subset N$ such that

$$\sum_{i \in N} b^i \sum_{S \subset N : i \in S} \lambda(S) = \sum_{i \in N} b^i,$$
$$\sum_{S \subset N} \lambda(S) v_\Gamma(S) > v_\Gamma(N).$$

From the linear independence of $(b^i)_{i \in N}$ it follows that $\sum_{i \in N} b^i z_i = \sum_{i \in N} b^i$ has a unique solution $z_i = 1$ for all $i \in N$. This implies that $\sum_{S \subset N : i \in S} \lambda(S) = z_i = 1$ for all $i \in N$, and hence, that λ is a balanced map. The balancedness of (N, v_Γ) then yields the contradiction $\sum_{S \subset N} \lambda(S) v_\Gamma(S) \leq v_\Gamma(N)$. Conclusion, there exists a vector $\xi \in \mathbb{R}^R$ satisfying (7.8). □

Although linear independence is a sufficient condition to guarantee the existence of an appropriate valuation scheme of the resources, this result cannot be obtained without allowing for negative values, as the following example shows.

Example 7.5 Consider the stochastic linear production game presented in Example 7.3, but now with

$$A = \begin{bmatrix} 1 \\ 0 \end{bmatrix}, \; b^1 = \begin{bmatrix} 0.1 \\ 0 \end{bmatrix}, \text{ and } b^2 = \begin{bmatrix} 1 \\ 1 \end{bmatrix}.$$

Note that the second resource is not needed for production, so that the corresponding TU-game (N, v_Γ) is the same as in Example 7.3. Since the resource bundles are linearly independent, we can find values $\xi = (\xi_1, \xi_2)$ such that $(\xi^\top b^1, \xi^\top b^2) \in C(v_\Gamma)$. Recall that $\xi_1 \geq 1.98$ and that $\xi^\top b^1 + \xi^\top b^2 = 1.1\xi_1 + \xi_2 = v_\Gamma(\{1,2\}) = 1.99$. Since $\xi_2 = 1.99 - 1.1\xi_1 \leq -0.188$, the superfluous resource must have a negative value.

[1]The system $Ax \leq b$ has a solution x if and only if there exists no $y \geq 0$ such that $y^\top A = 0$ and $y^\top b < 0$.

7.2 Financial Games

The fact that retail prices are uncertain when individuals decide upon their production plan, is particularly applicable to investment funds. For at the time that people invest their capital in financial assets, they are not sure about the future value of this asset. To reduce the volatility of the total returns, they may prefer to participate in an investment fund to obtain a more diversified portfolio.

Besides reducing volatility, people may also benefit from cooperation if the return varies with the amount of capital invested. Bank deposits, for instance, usually earn a higher interest rate when more capital is deposited. These problems were introduced by Lemaire (1983) and further analyzed by Izquierdo and Rafels (1996) and Borm et al. (1999). The latter extended the former model by considering several periods so as to include term-dependent interest rates. In contrast to the model we present here, the three previous models abstract from risk bearing investments. Instead, they assume that the earned interest rates are known with certainty beforehand.

In the financial games we consider, there is a society $N \subset \mathbb{N}$ of risk averse expected utility maximizing agents, each having a utility function of the form $U_i(t) = \beta_i e^{-\alpha_i t}$ with $\beta_i < 0$ and $\alpha_i > 0$. Each agent has capital $\omega^i \in \mathbb{R}_+$ available for investments in several, infinitely divisible risky assets $M \subset \mathbb{N}$. For each asset $j \in M$, let $\pi_j \in \mathbb{R}_+$ denote the asset price, $\boldsymbol{R}_j \in L^1(\mathbb{R}_+)$ the random future value, and $q_j \in \mathbb{R}_+$ the quantity purchased. Then a portfolio $q \in \mathbb{R}^M_+$ of assets is feasible for coalition $S \subset N$ if they have sufficient capital at their disposal, that is if $\sum_{j \in M} \pi_j q_j \leq \sum_{i \in S} \omega_i$. Furthermore, the revenue of a portfolio $q \in \mathbb{R}^R_+$ equals $\sum_{j \in M} \boldsymbol{R}_j q_j$. Since agents are not forced to invest all their capital in risky assets, we must assume that there is a riskless asset $j_0 \in M$ that earns the risk-free interest rate $r \geq 0$. Furthermore, we may assume without loss of generality that $\pi_{j_0} = 1$.

Summarizing, a *financial game* $\Gamma = (N, \mathcal{V}, \{\succsim_i\}_{i \in N})$ is described by

$$\mathcal{V}(S) = \left\{ \sum_{j \in M} \boldsymbol{R}_j q_j \,\middle|\, \exists_{q \in \mathbb{R}^M_+} : \sum_{j \in M} \pi_j q_j \leq \sum_{i \in S} \omega_i \right\}, \tag{7.9}$$

for all $S \subset N$, where the preferences $\succsim_i$ are induced by U_i. The class of financial games is denoted by $FG(N)$.

It is a straightforward exercise to see that $FG(N) \subset SLP(N)$. For investing in a risky asset can be described by a simple production technology, that uses a single resource, i.e. capital, to produce several commodities, i.e. financial assets. The production technology is determined by the asset prices $(\pi_j)_{j \in M}$: the asset price π_j denotes how much of the resource capital is needed to produce one unit of asset $j \in M$. The prices at which produced output can be sold are the random returns $\boldsymbol{R}_j$. Hence, we obtain the following result.

Theorem 7.6 Each financial game $\Gamma \in FG(N)$ is totally balanced.

7.2.1 The Optimal Investment Portfolio

The relatively simple production technology in financial games enables us to provide a more detailed analysis of the optimal investment portfolio. Using the Pareto optimality result of Proposition 7.1, the optimal portfolio q^* of coalition S is the optimal solution of

$$\begin{array}{rcll} \max\ \alpha(S) \sum_{j \in M} \log \left(\int_0^\infty e^{-q_j t/\alpha(S)} dF_{\boldsymbol{R}_j}(t) \right)^{-1} & & & \\ \text{s.t.: } \sum_{j\in M} \pi_j q_j & \leq & \sum_{i\in S} \omega^i, & \\ q_j & \geq & 0, & \text{for all } j \in M. \end{array} \tag{7.10}$$

Recall that there is a risk-free asset j_0 yielding the risk-free interest rate $r \geq 0$. Hence, (7.10) is equivalent to

$$\begin{array}{rcll} \max\ (1+r) q_{j_0} + \alpha(S) \sum_{j \in M_0} \log \left(\int_0^\infty e^{-q_j t/\alpha(S)} dF_{\boldsymbol{R}_j}(t) \right)^{-1} & & & \\ \text{s.t.: } \sum_{j\in M} \pi_j q_j & \leq & \sum_{i\in S} \omega^i, & \\ q_j & \geq & 0, & \text{for all } j \in M, \end{array}$$

where $M_0 = M \backslash \{j_0\}$. Instead of focusing on the quantities, we can also focus on the capital invested in each asset. For this purpose, let $w_j = \pi_j q_j$ denote the capital invested in asset $j \in M$. Then substituting $q_j = w_j/\pi_j$ yields

$$\begin{array}{rcll} \max\ (1+r) w_{j_0} - \alpha(S) \sum_{j \in M_0} \log \left(\int_0^\infty e^{-w_j/\pi_j \cdot t/\alpha(S)} dF_{\boldsymbol{R}_j}(t) \right) & & & \\ \text{s.t.: } \sum_{j\in M} w_j & \leq & \sum_{i\in S} \omega^i, & \\ w_j & \geq & 0, & \text{for all } j \in M. \end{array}$$

$$\begin{array}{rcll} = \quad \max\ (1+r) w_{j_0} + \sum_{j \in M_0} \int_0^{w_j} \frac{1}{\pi_j} \frac{\int_0^\infty t e^{-x_j/\pi_j \cdot t/\alpha(S)} dF_{\boldsymbol{R}_j}(t)}{\int_0^\infty e^{-x_j/\pi_j \cdot t/\alpha(S)} dF_{\boldsymbol{R}_j}(t)} dx_j & & & \\ \text{s.t.: } \sum_{j\in M} w_j & \leq & \sum_{i\in S} \omega_i, & \\ w_j & \geq & 0, & \text{for all } j \in M. \end{array}$$

Since coalition S prefers investing in the risk-free asset to not investing at all, all the capital is spent in the optimal portfolio. Thus, we may replace $\sum_{j\in M} w_j^* \leq \sum_{i\in S} \omega^i$ by $\sum_{j\in M} w_j = \sum_{i\in S} \omega^i$. Substituting $w_{j_0} = \sum_{i\in S} \omega^i - \sum_{j\in M_0} w_j$ gives

$$\begin{aligned}
& \max\ (1+r)\left(\sum_{i\in S}\omega^i - \sum_{j\in M_0} w_j\right) \\
& \qquad\qquad + \sum_{j\in M_0}\int_0^{w_j} \frac{1}{\pi_j}\frac{\int_0^\infty te^{-x_j/\pi_j\cdot t/\alpha(S)}dF_{\boldsymbol{R}_j}(t)}{\int_0^\infty e^{-x_j/\pi_j\cdot t/\alpha(S)}dF_{\boldsymbol{R}_j}(t)}dx_j \\
& \text{s.t.: } \textstyle\sum_{j\in M_0} w_j \ \le\ \sum_{i\in S}\omega^i, \\
& \qquad\qquad w_j \ \ge\ 0, \qquad\qquad \text{for all } j\in M_0. \\
= \quad & \max\ (1+r)\sum_{i\in S}\omega^i \\
& \quad + \sum_{j\in M_0}\int_0^{w_j} \frac{1}{\pi_j}\frac{\int_0^\infty te^{-x_j/\pi_j\cdot t/\alpha(S)}dF_{\boldsymbol{R}_j}(t)}{\int_0^\infty e^{-x_j/\pi_j\cdot t/\alpha(S)}dF_{\boldsymbol{R}_j}(t)} - (1+r)dx_j \\
& \text{s.t.: } \textstyle\sum_{j\in M_0} w_j \ \le\ \sum_{i\in S}\omega^i, \\
& \qquad\qquad w_j \ \ge\ 0, \qquad\qquad \text{for all } j\in M_0. \\
= \quad & \max\ (1+r)\sum_{i\in S}\omega^i + \sum_{j\in M_0}\int_0^{w_j}(MRI_j(x_j)-r)dx_j \qquad (7.11) \\
& \text{s.t.: } \textstyle\sum_{j\in M_0} w_j \ \le\ \sum_{i\in S}\omega^i, \\
& \qquad\qquad w_j \ \ge\ 0, \qquad\qquad \text{for all } j\in M_0,
\end{aligned}$$

where

$$MRI_j(x_j) = \frac{\dfrac{\int_0^\infty te^{-x_j/\pi_j\cdot t/\alpha(S)}dF_{\boldsymbol{R}_j}(t)}{\int_0^\infty e^{-x_j/\pi_j\cdot t/\alpha(S)}dF_{\boldsymbol{R}_j}(t)} - \pi_j}{\pi_j}. \qquad (7.12)$$

$MRI_j(x_j)$ represents the marginal return on investment in terms of certainty e-quivalents that coalition S receives from investing an additional dollar in the risky asset $j \in M_0$, when they have already invested x_j dollars in asset j. Note that the marginal return is decreasing[2] in the capital invested. Similarly, r represents the cost of capital, that is the marginal return of investing an additional dollar in the risk-free asset. In order to maximize the the total return, a coalition should invest in the asset with the largest marginal return on investment that exceeds the cost of capital. If such investments do not exist, the remaining capital should be invested in the risk-free asset. Hence, we can construct the optimal portfolio in the following way.

When no investments have been made, that is $x_j = 0$ for all $j \in M_0$, the marginal return of asset $j \in M_0$ equals

$$MRI_j(0) = \frac{E(\boldsymbol{R}_j) - \pi_j}{\pi_j}.$$

Assuming that $MRI_{j_1}(0) > MRI_{j_2}(0) > \ldots > MRI_{j_m}(0)$, it is optimal to invest in asset j_1. As the capital x_{j_1} invested in asset j_1 increases, the marginal

[2]This follows indirectly from Lemma 6.6 with $c = \frac{-1}{\pi_j\alpha(S)}$.

return $MRI_{j_1}(x_{j_1})$ decreases, and either one of the following three things will happen: the capital runs out, the marginal return becomes equal to the cost of capital, or the marginal benefit becomes equal to $MRI_{j_2}(0)$. In the first case, the current investments make up the optimal portfolio, while in the second case, the remaining capital is invested in the risk-free asset. In the third case, it is optimal to start investing in both asset j_1 and asset j_2. The quantities x_{j_1} and x_{j_2}, however, must be chosen such that the marginal returns of both assets remain equal to each other, that is $MRI_{j_1}(x_{j_1}) = MRI_{j_2}(x_{j_2})$. Again, as the investments increase, the marginal returns decrease simultaneously, and either of the following three things will happen: the capital runs out, the marginal return of both assets becomes equal to the cost of capital, or the marginal return becomes equal to $MRI_{j_3}(0)$. The procedure continues in a similar way until either the capital runs out, or the marginal return of all assets in the portfolio equals the cost of capital.

Let us illustrate this procedure with an example.

Example 7.7 Consider two individuals with $\alpha_1 = 0.1$ and $\alpha_2 = 0.05$, who can invest their individual capital of \$ 2 and \$ 4 in three risky assets and a risk-free asset. The risk-free interest rate is 10%. The future value of the risky assets is exponentially distributed with expected return of \$ 4, \$ 6, and \$ 8.1, respectively, while the current asset prices are \$ 3, \$ 5, and \$ 8, respectively. The marginal return on investment for asset j equals

$$MRI_j(x_j) = \frac{\alpha(S)E(\boldsymbol{R}_j)}{\alpha(S)\pi_j + x_j E(\boldsymbol{R}_j)} - 1$$

for all $x_j \geq 0$. The optimal portfolio for coalition $S = \{1, 2\}$ is now constructed as follows. When $x_j = 0$ for $j = 1, 2, 3$ we have that

$$\begin{aligned} MRI_1(0) &= \tfrac{E(\boldsymbol{R}_1)}{\pi_1} - 1 &= 0.333, \\ MRI_2(0) &= \tfrac{E(\boldsymbol{R}_2)}{\pi_2} - 1 &= 0.200, \\ MRI_3(0) &= \tfrac{E(\boldsymbol{R}_3)}{\pi_3} - 1 &= 0.013. \end{aligned}$$

Since asset 1 has the largest marginal return, coalition S starts with investing in asset 1 and increases this investment until $MRI_1(x_1) = MRI_2(0)$, which happens at $x_1 = 2.500$. They continue by investing in asset 1 and asset 2, such that $MRI_1(x_1) = MRI_2(x_2)$. This holds true if

$$x_1 = \alpha(S)\left(\frac{\pi_2}{E(\boldsymbol{R}_2)} - \frac{\pi_1}{E(\boldsymbol{R}_1)}\right) + x_2.$$

They increase the investments x_1 and x_2 until the available capital of \$ 6 is used up. Then $x_1 = 4.250$ and $x_2 = 1.750$. Note that for the current portfolio, the marginal return on investment $MRI_1(4.250) = MRI_2(1.750) = 0.1215$ still exceeds the cost of capital $r = 0.100$. Thus, coalition S would like to invest more in asset j_1 and asset j_2, but is constrained by its capital budget. Then the optimal investment portfolio equals $w^* = (4.250, 1.750, 0)$, or $q^* = (1.417, 0.350, 0)$ if stated in quantities.

Proposition 7.8 Let $\Gamma \in FG(N)$ be a financial game and let $q^* \in \mathbb{R}_+^{M_0}$ be the optimal portfolio for coalition $S \subset N$. Then
(a) $q_j^* = 0$ if $\frac{E(\boldsymbol{R}_j)-\pi_j}{\pi_j} \leq r$.
(b) $q_j^* > 0$ if there exists $k \in M_0$ such that $q_k^* > 0$ and $\frac{E(\boldsymbol{R}_j)-\pi_j}{\pi_j} > \frac{E(\boldsymbol{R}_k)-\pi_k}{\pi_k}$.

PROOF: Let $\Gamma \in FG(N)$ and $S \subset N$. From (7.11) it follows that coalition S does certainly not invest any capital in asset j if $MRI_j(x_j) \leq r$ for all $x_j \geq 0$. Since the marginal return is strictly decreasing in x_j, we obtain that $w_j^* = 0$ if $MRI_j(0) < r$. Hence, $q_j^* = \frac{w_j^*}{\pi_j} = 0$ if $MRI_j(0) < r$, which proves (1).

In order to prove (2), let $MRI_{j_1}(0) > MRI_{j_2}(0) > \ldots > MRI_{j_m}(0)$. Recall that in the construction of the optimal portfolio, coalition S starts with investing in asset j_1, continues with investing in asset j_1 and asset j_2, and so on. Thus, if $MRI_j(0) > MRI_k(0)$ and $w_k^* > 0$, then coalition S must also invest in asset j. Hence, $w_j^* > 0$ and, consequently, $q_j^* = \frac{w_j^*}{\pi_j} > 0$. □

Summarizing, a coalition will not invest in assets with an expected return on investment lower than the cost of capital, that is $\frac{E(\boldsymbol{R}_j)-\pi_j}{\pi_j} \leq r$. Furthermore, it is the asset's expected return on investment that determines whether or not a coalition should invest in it. The variability in the asset's return only plays a role in determining the quantity that is invested. Consider, for example, the following two assets. Asset 1 has a price of \$ 90 and yields \$ 1000 with probability 0.55 and \$ -1000 with probability 0.45. Asset 2 has a price of \$ 90 and yields \$ 99 with certainty. Since the expected return of asset 1 exceeds the expected return of asset 2 by 1.11%, it is optimal to start investing in asset 1, although it involves (much) more risk than asset 2.

7.2.2 The Proportional Rule

The proportional rule is most commonly used by investment funds to allocate the returns to the participants in the fund. This means that if an investment fund generates a rate of return of 12%, each individual participant earns 12% on the amount of capital that he contributed to the fund. So, the individual rate of return does not depend on the amount of capital brought in. Izquierdo and Rafels (1996) and Borm et al. (1999) show that the proportional rule results in a core-allocation for deposit games, in which the rate of return is risk-free and dependent on the term and the amount of the bank deposit. For the class of financial games introduced in this section, however, the proportional allocation-rule does not perform as well.

Let $\Gamma \in FG(N)$ be a financial game and let $q^N \in \mathbb{R}_+^M$ denote the optimal portfolio. We start with considering the proportional rule for the corresponding TU-game (N, v_Γ). In terms of certainty equivalents, the rate of return of the optimal portfolio q^N equals

$$\frac{v_\Gamma(N) - \omega(N)}{\omega(N)},$$

where $\omega(N) = \sum_{k \in N} \omega_k$. The proportional rule ρ is then defined by

$$\rho_i = \omega_i \left(1 + \frac{v_\Gamma(N) - \omega(N)}{\omega(N)}\right) = \frac{\omega_i}{\omega(N)} v_\Gamma(N), \tag{7.13}$$

for all $i \in N$.

The proportional rule ρ, however, is not proportional in the sense that each individual $i \in N$ receives the random payoff $\frac{\omega_i}{\omega(N)} \sum_{j \in N} \boldsymbol{R}_j q_j^N$. Since we considered the TU-game (N, v_Γ), the total return $\sum_{j \in M} \boldsymbol{R}_j q_j^N$ is allocated Pareto optimally instead of proportionally, that is

$$r_i = \frac{\frac{1}{\alpha_i}}{\sum_{k \in N} \frac{1}{\alpha_k}}$$

instead of $r_i = \frac{\omega_i}{\omega(N)}$, for all $i \in N$. It is the certainty equivalent of this Pareto optimal allocation that is distributed proportionally by ρ. Note, however, that

$$r_i = \frac{\frac{1}{\alpha_i}}{\sum_{k \in N} \frac{1}{\alpha_k}} = \frac{\omega_i}{\omega(N)}$$

if and only if $\frac{\omega_i}{\alpha(\{i\})} = \frac{\omega_k}{\alpha(\{k\})}$ for all $i, k \in N$.

Theorem 7.9 Let $\Gamma \in FG(N)$. Then $\rho \in C(v_\Gamma)$ if and only if $\frac{\omega_i}{\alpha(\{i\})} = \frac{\omega_k}{\alpha(\{k\})}$ for all $i, k \in N$.

PROOF: Take $\Gamma \in FG(N)$. For all $S \subset N \mathbb{R}_+^M$ denote the optimal portfolio and let $\omega(S) = \sum_{i \in S} \omega_i$. If $\frac{\omega_i}{\alpha(\{i\})} = \frac{\omega_k}{\alpha(\{k\})}$ for all $i, k \in N$, then for all $S, T \subset N$ it holds true that

$$\frac{\omega(S)}{\alpha(S)} = \sum_{i \in S} \frac{\alpha(\{i\})}{\alpha(S)} \frac{\omega_i}{\alpha(\{i\})} = \sum_{k \in S} \frac{\alpha(\{k\})}{\alpha(T)} \frac{\omega_k}{\alpha(\{k\})} = \frac{\omega(T)}{\alpha(T)}.$$

For each $S \subset N$ we have that

$$
\begin{aligned}
\sum_{i \in S} \rho_i &= \frac{\omega(S)}{\omega(N)} \alpha(N) \sum_{j \in M} \log \left(\int_0^\infty e^{-q_j^N t/\alpha(N)} dF_{\boldsymbol{R}_j}(t) \right)^{-1} \\
&\geq \frac{\omega(S)}{\omega(N)} \alpha(N) \sum_{j \in M} \log \left(\int_0^\infty e^{-\frac{\omega(N)}{\omega(S)} q_j^S t/\alpha(N)} dF_{\boldsymbol{R}_j}(t) \right)^{-1} \\
&= \alpha(S) \sum_{j \in M} \log \left(\int_0^\infty e^{-q_j^S t/\alpha(S)} dF_{\boldsymbol{R}_j}(t) \right)^{-1} \\
&= v_\Gamma(S),
\end{aligned}
$$

where the inequality follows from the fact that $\left(\frac{\omega(N)}{\omega(S)} q_j^S\right)_{j \in M}$ is a feasible portfolio for coalition N, and the second equality follows from $\frac{\omega(S)}{\omega(N)}\alpha(N) = \alpha(S)$. Hence, $\rho \in C(v_\Gamma)$.

Next, suppose that $\frac{\omega_{k_1}}{\alpha(\{k_1\})} < \frac{\omega_{k_2}}{\alpha(\{k_2\})}$ for some $k_1, k_2 \in N$. Since

$$
\frac{\omega_{k_1}}{\alpha(\{k_2\})} < \frac{\omega(N)}{\alpha(N)} < \frac{\omega_{k_2}}{\alpha(\{k_2\})}
$$

there exists $S \subset N$ such that $\frac{\omega(S)}{\alpha(S)} < \frac{\omega(N)}{\alpha(N)}$. Then

$$
\begin{aligned}
\sum_{i \in S} \rho_i &= \alpha(S) \sum_{i \in S} \frac{\rho_i}{\alpha(S)} \\
&= \alpha(S) \frac{\omega(S)\alpha(N)}{\omega(N)\alpha(S)} \sum_{j \in M} \log \left(\int_0^\infty e^{-q_j^N t/\alpha(N)} dF_{\boldsymbol{R}_j}(t) \right)^{-1} \\
&< \alpha(S) \sum_{j \in M} \log \left(\int_0^\infty e^{-\frac{\omega(S)\alpha(N)}{\omega(N)\alpha(S)} q_j^N t/\alpha(N)} dF_{\boldsymbol{R}_j}(t) \right)^{-1} \\
&= \alpha(S) \sum_{j \in M} \log \left(\int_0^\infty e^{-\frac{\omega(S)}{\omega(N)} q_j^N t/\alpha(S)} dF_{\boldsymbol{R}_j}(t) \right)^{-1} \\
&\leq \alpha(S) \sum_{j \in M} \log \left(\int_0^\infty e^{-q_j^S t/\alpha(S)} dF_{\boldsymbol{R}_j}(t) \right)^{-1} \\
&= v_\Gamma(S),
\end{aligned}
$$

where the first inequality follows from Jensen's inequality, $\frac{\omega(S)\alpha(N)}{\omega(N)\alpha(S)} < 1$, and $\boldsymbol{R}_j$ nondegenerate for at least one $j \in M$. The second inequality follows from the fact that $\left(\frac{\omega(S)}{\omega(N)} q_j^S\right)_{j \in M}$ is a feasible portfolio for coalition S. Hence, $\rho \notin C(v_\Gamma)$. □

7.3 Remarks

Similar to insurance games, the results presented in this chapter still go through if we replace mutual independency of $(\boldsymbol{P}_j)_{j \in M}$ by the assumption that the covariance matrix of the retail prices $(\boldsymbol{P}_j)_{j \in M}$ is negative definite.

In this chapter we only discussed price uncertainty, but the stochastic linear production game also applies to situations where there is uncertainty in the production process due to production losses. Production losses may occur when produced output does not satisfy prespecified (quality) standards. We can express these losses as a percentage of the production plan. Consider, for instance, a producer of audio and video tapes, and suppose that production losses are 2% for audio tapes and 0.7% for video tapes. Then given a feasible production plan $c \in \mathbb{R}^2_+$, the proceeds are $0.98p_a c_a + 0.993p_v c_v$.

Fall out of production, however, is generally uncertain at the start of production. Let $1 - \boldsymbol{X}^i_j$ denote the stochastic percentage of production losses of commodity j when using technology i, and let p_j denote the deterministic retail price of good $j \in M$. Given a feasible production plan $c \in C(S)$, coalition $S \subset N$ obtains the payoff $\sum_{j\in M}\sum_{i\in S} p_j \boldsymbol{X}^i_j c^i_j$. The corresponding stochastic cooperative game $(N, \mathcal{V}, (\succsim_i)_{i\in N})$ is given by

$$\mathcal{V}(S) = \left\{ \sum_{j\in M}\sum_{i\in S} p_j \boldsymbol{X}^i_j c^i_j \,\middle|\, \sum_{i\in S} c^i \in C(S) \right\}. \tag{7.14}$$

for all $S \subset N$, and $\succsim_i$ induced by U_i for each $i \in N$. Note that this model can easily be written in terms of a stochastic linear production game by considering c^i_j and c^k_j, $i \neq k$, as two different commodities. Hence, they are totally balanced as well.

A Probability Theory

This appendix provides the reader with a brief introduction into probability theory. We will confine with explaining those terms that are needed to understand this monograph. Furthermore, we will refrain from any mathematical details on this subject. For an extensive discussion on measure theory and probability theory in particular we refer to Burrill (1972) or Feller (1950), (1966).

A stochastic variable describes a situation or experiment whose outcome is determined by chance. For instance, the number of points one obtains when throwing a die can be described by a stochastic variable. Before giving a general definition of a stochastic variable, let us elaborate on this example.

Example A.1 When throwing a fair die, the outcome can be any integer between 1 and 6. Let us denote these outcomes by the set Ω. So, $\Omega = \{1,2,3,4,5,6\}$. Obviously, one does not know the outcome of this experiment beforehand. What one does know, however, is the probability with which each outcome occurs. For instance, the chance that the experiment results in three points is $\frac{1}{6}$. In other words, the event that the die ends up with three dots faced upward occurs with probability $\frac{1}{6}$. This, of course, is only one particular event. We can think of many other events, for instance, the event that the number of points is even, or that the number of points is less than 4. Such events can be described by subsets of outcomes. For example, $E = \{2,4,6\} \subset \Omega$ describes the event that the outcome is an even number of points. Now, let $\mathcal{H}$ denote the set of all possible events. Thus,

$$\mathcal{H} = \{E \mid E \subset \Omega\}.$$

For each event $E \in \mathcal{H}$ we know the probability that this event occurs. We already noted that the event $E = \{3\}$ occurs with probability $\frac{1}{6}$. The probabilities with which these events occur are described by a function $\mathbb{P} : \mathcal{H} \to [0,1]$. For this case this implies that $\mathbb{P}(\{\omega\}) = \frac{1}{6}$ for all $\omega \in \Omega$. Furthermore, we have that

$$\mathbb{P}(\text{the outcome is even}) = \mathbb{P}(\{2,4,6\}) = \tfrac{1}{2},$$

and

$$\mathbb{P}(\text{the outcome is less than } 4) = \mathbb{P}(\{1,2,3\}) = \tfrac{1}{2}.$$

Summarizing, the experiment of throwing a die is described by the triple $(\Omega, \mathcal{H}, \mathbb{P})$.

The triple $(\Omega, \mathcal{H}, \mathbb{P})$ in Example A.1 is called a probability space. In general, each chance experiment is described by a *probability space* $(\Omega, \mathcal{H}, \mathbb{P})$, where Ω is the outcome space, $\mathcal{H}$ is a σ-algebra on Ω, and $\mathbb{P}$ is a probability measure on the σ-algebra $\mathcal{H}$. The terms σ-algebra and probability measure need some further explanation. Let $\mathcal{H}$ be a set of events. Then $\mathcal{H}$ is called a *σ-algebra* on Ω if the following three conditions are satisfied:

(i) $\emptyset \in \mathcal{H}$,

(ii) if $E \in \mathcal{H}$ then $\Omega \backslash E \in \mathcal{H}$,

(iii) if $E_k \in \mathcal{H}$ for $k = 1, 2, \ldots$, then $\cup_{k=1}^{\infty} E_k \in \mathcal{H}$.

Example A.2 Let $\Omega = \{1, 2, 3, 4\}$. An example of a σ-algebra on Ω is

$$\mathcal{H} = \{\emptyset, \{2\}, \{3\}, \{2,3\}, \{1,4\}, \{1,3,4\}, \{1,2,4\}, \{1,2,3,4\}\}.$$

Example A.3 Let $\Omega = \mathbb{R}$. The smallest σ-algebra on $\mathbb{R}$ is $\mathcal{H}_1 = \{\emptyset, \mathbb{R}\}$. A σ-algebra that contains $\mathcal{H}_1$ is the σ- algebra $\mathcal{H}_2$ given by $\mathcal{H}_2 = \{\emptyset, A, \mathbb{R}\backslash A, \mathbb{R}\}$, with A any subset of $\mathbb{R}$. The largest σ-algebra on $\mathbb{R}$ one can construct consists of all subsets of $\mathbb{R}$, that is, $\mathcal{H}_3 = \{A |\ A \subset \mathbb{R}\}$. Note that $\mathcal{H}_1 \subset \mathcal{H}_2 \subset \mathcal{H}_3$.

A *probability measure* $\mathbb{P}$ is a function assigning to each event $E \in \mathcal{H}$ a nonnegative number such that

(i) $\mathbb{P}(\emptyset) = 0$ and $\mathbb{P}(\Omega) = 1$,

(ii) for any mutually disjoint events E_k, $k = 1, 2, \ldots$, it holds that $\mathbb{P}(\cup_{k=1}^{\infty} E_k) = \sum_{k=1}^{\infty} \mathbb{P}(E_k)$.

Condition (ii) implies that for any disjoint events E_1 and E_2, the probability that either of the two occurs equals the sum of the separate probabilities. For instance, consider in Example A.1 the events $\{3\}$ and $\{4\}$. The probability that one of the two events occurs equals $\mathbb{P}(\{3\} \cup \{4\}) = \mathbb{P}(\{3\}) + \mathbb{P}(\{4\}) = \frac{1}{6} + \frac{1}{6} = \frac{1}{3}$. Note that the events must be disjoint. For if we consider the events $\{3\}$ and $\{3,4\}$ then

$$\tfrac{1}{3} = \mathbb{P}(\{3,4\}) = \mathbb{P}(\{3\} \cup \{3,4\}) \neq \mathbb{P}(\{3\}) + \mathbb{P}(\{3,4\}) = \tfrac{1}{6} + \tfrac{1}{3} = \tfrac{1}{2}.$$

Let $(\Omega, \mathcal{H}, \mathbb{P})$ be a probability space and let $E_1, E_2 \in \mathcal{H}$. The events E_1 and E_2 are called *independent* if

$$\mathbb{P}(E_1 \cap E_2) = \mathbb{P}(E_1)\mathbb{P}(E_2). \tag{A.1}$$

Independence implies that occurrence of the event E_1 gives no additional information on the occurrence of the event E_2, and the other way around.

Example A.4 Consider the experiment described in Example A.1. The events $E_1 = \{2\}$ and $E_2 = \{2,4,6\}$ are not independent because $\frac{1}{6} = \mathbb{P}(E_1 \cap E_2) \neq \mathbb{P}(E_1)\mathbb{P}(E_2) = \frac{1}{6} \cdot \frac{1}{2} = \frac{1}{12}$. Indeed, if one knows that the event E_1 has occurred, then the outcome is even. Hence, one knows with certainty that the event E_2 has also occurred. Furthermore, if one knows that E_2 has occurred, that is, the outcome is even, then the probability that event E_1 occurs equals $\frac{1}{3}$ instead of $\frac{1}{6}$.

Now that we are familiar with probability spaces, we can introduce stochastic variables. For this purpose we need to introduce the Borel σ-algebra. The *Borel σ-algebra* $\mathcal{B}$ is the smallest σ-algebra on $\mathbb{R}$ containing all half open intervals $(a, b]$ with $-\infty < a < b < \infty$. Furthermore, an element $B \in \mathcal{B}$ is called a Borel-set. Exactly which sets the Borel σ-algebra contains is difficult to say. Besides all the half open intervals, it also contains all singletons $\{a\}$, $a \in \mathbb{R}$. Other examples of Borel sets are $(-\infty, b]$ with $b \in \mathbb{R}$, $[a, \infty)$ with $a \in \mathbb{R}$, and $[a, b]$ with $-\infty < a \leq b < \infty$.

A *stochastic variable* $\boldsymbol{X}$ is a measurable function assigning to each outcome $\omega \in \Omega$ a real number $\boldsymbol{X}(\omega)$. A function $\boldsymbol{X}$ is *measurable* with respect to $\mathcal{H}$ if for all $B \in \mathcal{B}$ it holds that $\{\omega \in \Omega | \boldsymbol{X}(\omega) \in B\} \in \mathcal{H}$. Measurability thus means that all outcomes ω that map into a Borel set B constitute an event $E \in \mathcal{H}$. Consequently, we can determine the probability that $\boldsymbol{X} \in B$ for every $B \in \mathcal{B}$. For ease of notation, the event $\{\omega \in \Omega | \boldsymbol{X}(\omega) \in B\}$ is often abbreviated to $\{\boldsymbol{X} \in B\}$.

Example A.5 Let us return to the experiment presented in Example A.1. The stochastic variable $\boldsymbol{X}$ defined by $\boldsymbol{X}(\omega) = \omega$ for all $\omega \in \Omega$ describes the experiment of throwing a die. So does the stochastic variable $\boldsymbol{Y}$ with $\boldsymbol{Y}(\omega) = 7 - \omega$ for all $\omega \in \Omega$. The difference, however, is in the interpretation of the realizations. If $x = 5$ is a realization of $\boldsymbol{X}$ this means that the outcome is five, but if $y = 5$ is a realization of $\boldsymbol{Y}$ then the outcome of the experiment is only two.

Let $(\Omega, \mathcal{H}, \mathbb{P})$ be a probability space and let $\boldsymbol{X}$ and $\boldsymbol{Y}$ be two stochastic variables on the outcome set Ω. Then $\boldsymbol{X}$ and $\boldsymbol{Y}$ are *independent* if for any two Borel sets $B_1, B_2 \in \mathcal{B}$ it holds that

$$\mathbb{P}((\boldsymbol{X}, \boldsymbol{Y}) \in B_1 \times B_2) = \mathbb{P}(\boldsymbol{X} \in B_1)\mathbb{P}(\boldsymbol{Y} \in B_2). \tag{A.2}$$

Independence of stochastic variables implies that the outcome of one stochastic variable provides no additional information on the outcome of another one.

A stochastic variable $\boldsymbol{X}$ is called *nonnegative* if $\boldsymbol{X}(\omega) \geq 0$ for all $\omega \in \Omega$ and is denoted by $\boldsymbol{X} \geq 0$. Similarly, we have that $\boldsymbol{X} \geq \boldsymbol{Y}$ if $\boldsymbol{X}(\omega) \geq \boldsymbol{Y}(\omega)$ for all $\omega \in \Omega$. Furthermore, if $\alpha, \beta \in \mathbb{R}$ we define the stochastic variable $\boldsymbol{Z} = \alpha \boldsymbol{X} + \beta \boldsymbol{Y}$ by $\boldsymbol{Z}(\omega) = \alpha \boldsymbol{X}(\omega) + \beta \boldsymbol{Y}(\omega)$ for all $\omega \in \Omega$.

Corresponding to each stochastic variable $\boldsymbol{X}$ we define the *probability distribution function* $F_{\boldsymbol{X}} : \mathbb{R} \to [0,1]$ by

$$F_{\boldsymbol{X}}(t) = \mathbb{P}(\{\boldsymbol{X} \leq t\}) = \mathbb{P}(\{\omega \in \Omega | \ \boldsymbol{X}(\omega) \leq t\}), \tag{A.3}$$

for all $t \in \mathbb{R}$. So, $F_{\boldsymbol{X}}(t)$ denotes the probability that the value of $\boldsymbol{X}$ is less than or equal to t. Note that for every probability disitribution function $F_{\boldsymbol{X}}$ it holds that $\lim_{t\to-\infty} F_{\boldsymbol{X}}(t) = 0$, $\lim_{t\to\infty} F_{\boldsymbol{X}}(t) = 1$, and that $F_{\boldsymbol{X}}$ is continuous from the right. Furthermore, probability distribution functions are (weakly) increasing and therefore discontinuous in at most countably many points.

Example A.6 Consider the stochastic variable $\boldsymbol{X}$ defined in Example A.5. The probability distribution function of $\boldsymbol{X}$ is illustrated in Figure A.1.

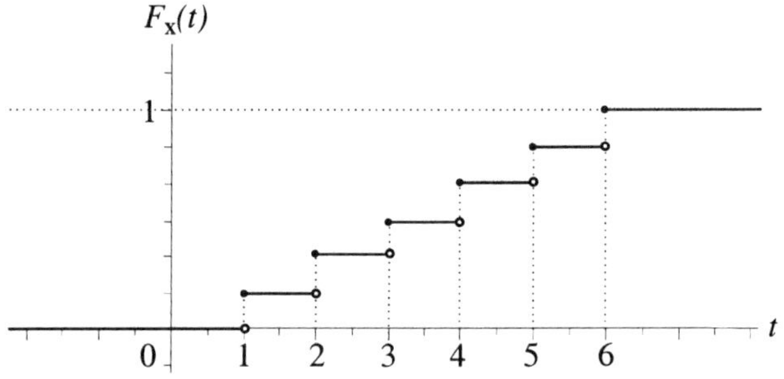

Figure A.1

A stochastic variable is uniquely determined by its probability distribution function. By this we mean the following. Given two probability spaces $(\Omega_1, \mathcal{H}_1, \mathbb{P}_1)$ and $(\Omega_2, \mathcal{H}_2, \mathbb{P}_2)$ and two stochastic variables $\boldsymbol{X} : \Omega_1 \to \mathbb{R}$ and $\boldsymbol{Y} : \Omega_2 \to \mathbb{R}$, we have that

$$\mathbb{P}_1(\{\omega \in \Omega_1 | \ \boldsymbol{X}(\omega) \in B\}) = \mathbb{P}_2(\{\omega \in \Omega_2 | \ \boldsymbol{Y}(\omega) \in B\})$$

for all $B \subset \mathcal{B}$ if and only if

$$F_{\boldsymbol{X}}(t) = F_{\boldsymbol{Y}}(t)$$

for all $t \in \mathbb{R}$. So, when working with stochastic variables, it suffices to know the distribution function; as a consequence, the underlying probability space is often left unspecified.

A stochastic variable $\boldsymbol{X}$ is called *degenerate* if it attains exactly one value with positive probability. So, there exists $y \in \mathbb{R}$ so that $\mathbb{P}(\{\boldsymbol{X} = x\} = 1$ if $x = y$ and $\mathbb{P}(\{\boldsymbol{X} = x\} = 0$ if $x \neq y$. In particular, we denote by $\mathbf{1}$ the degenerate stochastic variables with value 1. For ease of notation, we will use both x and $x\mathbf{1}$ to denote the degenerate stochastic variable $x\mathbf{1}$.

A stochastic variable $\boldsymbol{X}$ is called *discrete* if it can attain only countably many values. This means that there exists numbers $x_k \in \mathbb{R}$ and $p_k \in [0,1]$, $k = 1, 2, \ldots$, such that $\mathbb{P}(\{\boldsymbol{X} = x_k\}) = p_k$ for $k = 1, 2, \ldots$, and $\sum_{k=1}^{\infty} p_k = 1$. Note that for any Borel-set $B \in \mathcal{B}$ it holds that

$$\mathbb{P}(\{\boldsymbol{X} \in B\}) = \sum_{k \in \mathbb{N}: x_k \in B} p_k.$$

The stochastic variable $\boldsymbol{X}$, introduced in Example A.5 to describe the outcome of throwing a die, is an example of a discrete stochastic variable. $\boldsymbol{X}$ only attains six different values, namely, $x_k = k$, $k = 1, 2, \ldots, 6$. Furthermore, we have that $p_k = \frac{1}{6}$ for $k = 1, 2, \ldots, 6$.

A stochastic variable $\boldsymbol{X}$ is called *continuous* if there exists a continuous function $f_{\boldsymbol{X}} : \mathbb{R} \to \mathbb{R}$ such that

$$F_{\boldsymbol{X}}(t) = \int_{\infty}^{t} f_{\boldsymbol{X}}(t)dt, \tag{A.4}$$

for all $t \in \mathbb{R}$. The function $f_{\boldsymbol{X}}$ is called the *density function* of the stochastic variable $\boldsymbol{X}$. For any interval $[a, b]$ it then holds that

$$\mathbb{P}(\{\boldsymbol{X} \in [a, b]\}) = \int_{a}^{b} f_{\boldsymbol{X}}(t)dt.$$

Note that in the expression above we can replace the interval $[a, b]$ by the intervals $(a, b]$, $[a, b)$, and (a, b). Examples of continuous stochastic variables are provided at the end of this appendix.

When throwing a die for numerous times in a row, the average of all the outcomes will be around $3\frac{1}{2}$, that is, the average of the numbers 1 through 6. Therefore, we say that the expectation of the outcome equals $3\frac{1}{2}$. If $\boldsymbol{X}$ is a discrete stochastic variable, the *expectation* of $\boldsymbol{X}$ is given by

$$E(\boldsymbol{X}) = \sum_{k=1}^{\infty} p_k x_k.$$

If $\boldsymbol{X}$ is continuous with density function $f_{\boldsymbol{X}}$, the *expectation* is given by

$$E(\boldsymbol{X}) = \int_{-\infty}^{\infty} t f_{\boldsymbol{X}}(t)dt.$$

The following theorem is known as Jensen's inequality.

Theorem A.7 Let $\boldsymbol{X} \in L^1(\mathbb{R})$ be a random variable and let $g : \mathbb{R} \to \mathbb{R}$ be a convex function such that $E(\boldsymbol{X})$ and $E(g(\boldsymbol{X}))$ exist. Then $g(E(\boldsymbol{X})) \leq E(g(\boldsymbol{X}))$. In particular, if g is strict convex, the inequality is binding if and only if $\boldsymbol{X}$ is degenerate.

Example A.8 Consider the following two lotteries $\boldsymbol{X}$ and $\boldsymbol{Y}$ given by

$$\boldsymbol{X} = \begin{cases} 0, & \text{with probability } \frac{1}{2} \\ 20,000, & \text{with probability } \frac{1}{2} \end{cases}$$

and

$$\boldsymbol{Y} = \begin{cases} -10,000, & \text{with probability } \frac{9}{11} \\ 100,000, & \text{with probability } \frac{2}{11} \end{cases}$$

So when playing the lottery $\boldsymbol{X}$, there is a 50% chance of winning \$ $20,000$. But when playing the lottery $\boldsymbol{Y}$ there is about 18% change of winning no less than \$ $100,000$. On the darker side though, there is also about 82% chance that one has to pay \$ $10,000$. Both lotteries, however, have the same expected value of \$ $10,000$.

Although both lotteries defined in Example A.8 yield the same expected revenue, most people will prefer the lottery $\boldsymbol{X}$ to the lottery $\boldsymbol{Y}$. The risk of paying \$ $10,000$ in the second lottery is too high compared to \$ $100,000$ one can win. So, when choosing between two different lotteries, not only the expected value plays a role, but also the variation in the different outcomes. A measure for the variation in the outcomes of a stochastic variable $\boldsymbol{X}$ is the *variance* $V(\boldsymbol{X})$. For discrete random variables $\boldsymbol{X}$ the variance is defined as

$$V(\boldsymbol{X}) = \sum_{k=1}^{\infty} p_k (x_k - E(\boldsymbol{X}))^2,$$

while for continuous stochastic variables with density function $f_{\boldsymbol{X}}$ it is defined as

$$V(\boldsymbol{X}) = \int_{-\infty}^{\infty} (t - E(\boldsymbol{X}))^2 f_{\boldsymbol{X}}(t) dt.$$

For the lotteries defined in Example A.8 we thus have that

$$V(\boldsymbol{X}) = \tfrac{1}{2}(10,000)^2 + \tfrac{1}{2}(10,000)^2 = 1 \cdot 10^8,$$

and

$$V(\boldsymbol{Y}) = \tfrac{9}{11}(-20,000)^2 + \tfrac{2}{11}(90,000)^2 = 17 \cdot 10^8.$$

As the variance shows, the outcomes for the lottery $\boldsymbol{Y}$ vary much more than for the lotter $\boldsymbol{X}$.

Other characteristic values of stochastic variables are quantiles. For example, the 10%-quantile of a stochastic variable $\boldsymbol{X}$ is the largest value ξ such that the outcome of $\boldsymbol{X}$ will be less than ξ with at most 10% chance. Of course, the quantile can be determined for any percentage between 0 and 100. Formally, the *α-quantile* of a stochastic variable $\boldsymbol{X}$ is defined by

$$\xi_\alpha(\boldsymbol{X}) = \sup\{t \mid \mathbb{P}(\boldsymbol{X} < t) \leq \alpha\}, \tag{A.5}$$

where $\alpha \in (0,1)$. In particular, the 0.5-quantile is called the *median* of $\boldsymbol{X}$. Figure A.2 shows how quantiles can be derived from the graph of the probability distribution function.

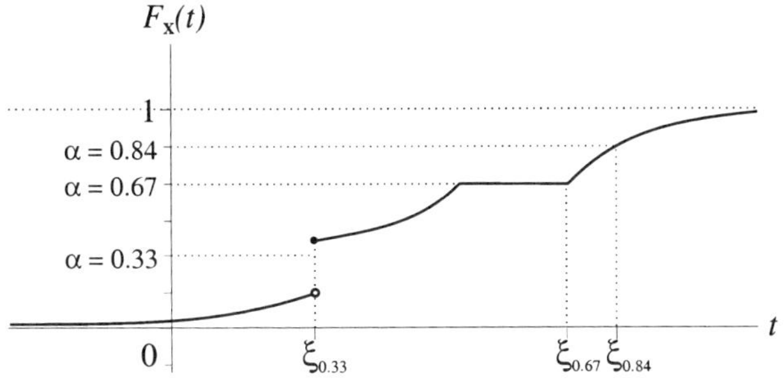

Figure A.2

Let $L^1(\mathbb{R})$ denote the set of all real valued stochastic variables with finite expectation, that is, $E(|\boldsymbol{X}|) < \infty$. Let $F_{\boldsymbol{X}}$ denote the probability distribution function of $\boldsymbol{X} \in L^1(\mathbb{R})$. Furthermore, define $\mathcal{F} = \{F_{\boldsymbol{X}} \mid \boldsymbol{X} \in L^1(\mathbb{R})\}$. Then $(\mathcal{F}, \rho)$ with

$$\rho(F,G) = \int_{-\infty}^{\infty} |F(t) - G(t)|e^{-|t|}dt \tag{A.6}$$

for all $F, G \in \mathcal{F}$ is a metric space. Next, let $(F_k)_{k\in\mathbb{N}}$ be a sequence in $\mathcal{F}$. Then the sequence $(F_k)_{k\in\mathbb{N}}$ *weakly converges* to $F \in \mathcal{F}$, denoted by $F_k \xrightarrow{w} F$, if $\lim_{k\to\infty} F_k(t) = F(t)$ for all $t \in \{t' \in \mathbb{R} \mid F \text{ is continuous in } t'\}$. Note that since probability distribution functions are continuous from the right, this limit is unique. Moreover, we have that $F_{\boldsymbol{X}_k} \xrightarrow{w} F_{\boldsymbol{X}}$ if and only if $\lim_{k\to\infty} \rho(F_{\boldsymbol{X}_k}, F_{\boldsymbol{X}}) = 0$.

We say that a sequence $(\boldsymbol{X}_k)_{k\in\mathbb{N}}$ of random variables in $L^1(\mathbb{R})$ converges to the random variable $\boldsymbol{X} \in L^1(\mathbb{R})$ if and only if the corresponding sequence

$(F_{\boldsymbol{X}_k})_{k\in\mathbb{N}}$ of probability distribution functions weakly converges to the probability distribution function $F_{\boldsymbol{X}}$ of $\boldsymbol{X}$. The following theorem is known in the literature as one of the *Helly theorems*.

Theorem A.9 Let $(\boldsymbol{X}^k)_{k\in\mathbb{N}}$ be a sequence in $L^1(\mathbb{R})$ weakly converging to $\boldsymbol{X} \in L^1(\mathbb{R})$, i.e., $F_{\boldsymbol{X}^k} \xrightarrow{w} F_{\boldsymbol{X}}$. If $g : \mathbb{R} \to \mathbb{R}$ is a continuous and bounded function then $\lim_{k\to\infty} E(g(\boldsymbol{X}^k)) = E(g(\boldsymbol{X}))$.

References

Alegre, A. and Mercè Claramunt, M.: 1995, Allocation of solvency cost in group annuities: actuarial principles and cooperative game theory, *Insurance: Mathematics & Economics* **17**, 19–34.

Anand, L., Shashishekhar, N., Ghose, D. and Prasad, U.: 1995, A survey of solution concepts in multicriteria games, *Journal of the Indian Institute of Science* **75**, 141–174.

Aumann, R.: 1960, Linearity of unrestricted transferable utilities, *Naval Research Logistics Quarterly* **7**, 281–284.

Aumann, R. and Peleg, B.: 1960, Von neumann-morgenstern solutions to cooperative games without sidepayments, *Bulletin of the American Mathematical Society* **66**, 173–179.

Bergstresser, K. and Yu, P.: 1977, Domination structures and multi-criteria problems in n-person games, *Theory and Decision* **8**, 5–48.

Bondareva, O.: 1963, Some applications of linear programming methods to the theory of cooperative games, *Problemi Kibernet* **10**, 119–139. In Russian.

Borch, K.: 1962a, Equilibrium in a reinsurance market, *Econometrica* **30**, 424–444.

Borch, K.: 1962b, Application of game theory to some problems in automobile insurance, *Astin Bulletin* **2**, 208–221.

Borm, P., De Waegenaere, A., Rafels, A., Suijs, J., Tijs, S. and Timmer, J.: 1999, Cooperation in capital deposits, *CentER Discussion Paper 9931*, Tilburg University.

Bühlmann, H.: 1980, An economic premium principle, *Astin Bulletin* **11**, 52–60.

Bühlmann, H.: 1984, The general economic premium principle, *Astin Bulletin* **14**, 13–21.

Burrill, C.: 1972, *Measure, Integration, and Probability*, McGraw-Hill, New York.

Charnes, A. and Granot, D.: 1973, Prior solutions: extensions of convex nucleolus solutions to chance-constrained games, *Proceedings of the Computer Science and Statistics Seventh Symposium at Iowa State University*, pp. 323–332.

Charnes, A. and Granot, D.: 1976, Coalitional and chance-constrained solutions to n-person games I, *SIAM Journal on Applied Mathematics* **31**, 358–367.

Charnes, A. and Granot, D.: 1977, Coalitional and chance-constrained solutions to n-person games II, *Operations Research* **25**, 1013–1019.

Curiel, I., Pederzoli, G. and Tijs, S.: 1989, Sequencing games, *European Journal of Operational Research* **40**, 344–351.

Debreu, G.: 1959, *Theory of Value*, Yale University Press, New Haven.

Eeckhoudt, L. and Gollier, C.: 1995, *Risk: Evaluation, Management, and Sharing*, Harvester Wheatsheaf, New York.

Feller, W.: 1950, *An Introduction to Probability Theory and its Applications*, Vol. 1, Wiley, New York.

Feller, W.: 1966, *An Introduction to Probability Theory and its Applications*, Vol. 2, Wiley, New York.

Goovaerts, M., De Vylder, F. and Haezendonck, J.: 1984, *Insurance Premiums*, North Holland, Amsterdam.

Granot, D.: 1977, Cooperative games in stochastic characteristic function form, *Management Science* **23**, 621–630.

Hamers, H.: 1996, *Sequencing and Delivery Situations: a Game Theoretic Approach*, Center dissertation series, Tilburg University.

Hamers, H., Suijs, J., Borm, P. and Tijs, S.: 1996, The split core for sequencing games, *Games and Economic Behavior* **15**, 165–176.

Harsanyi, J.: 1963, A simplified bargaining model for the n-person cooperative game, *International Economic Review* **4**, 194–220.

Hsiao, C. and Raghavan, T.: 1993, Shapley value for multi-choice cooperative games, *Games and Economic Behavior* **5**, 240–256.

Ichiishi, T.: 1981, Supermodularity: applications to convex games and the greedy algorithm for LP, *Journal of Economic Theory* **25**, 283–286.

Izquierdo, J. and Rafels, C.: 1996, A generalization of the bankruptcy game: financial cooperative games, *Working Paper E96/09*, University of Barcelona.

Kalai, E. and Samet, D.: 1985, Monotonic solutions to general cooperative games, *Econometrica* **53**, 307–327.

Kaneko, M.: 1976, On transferable utility, *International Journal of Game Theory* **5**, 183–185.

Lemaire, J.: 1983, An application of game theory: cost allocation, *Astin Bulletin* **14**, 61–81.

Lemaire, J.: 1991, Cooperative game theory and its insurance applications, *Astin Bulletin* **21**, 17–40.

Littlechild, S.: 1974, A simple expression for the nucleolus in a special case, *International Journal of Game Theory* **3**, 21–29.

Magill, M. and Shafer, W.: 1991, Incomplete markets, *in* W. Hildenbrand and H. Sonnenschein (eds), *Handbook of Mathematical Economics*, Vol. 6, Elsevier Science Publishers, Amsterdam.

Maschler, M. and Owen, G.: 1989, The consistent Shapley value for hyperplane games, *International Journal of Game Theory* **18**, 389–407.

Maschler, M. and Owen, G.: 1992, The consistent Shapley value for games without side payments, *in* R. Selten (ed.), *Rational Interaction*, Springer-Verlag, Berlin.

Maschler, M., Potters, J. and Tijs, S.: 1992, The general nucleolus and the reduced game property, *International Journal of Game Theory* **21**, 85–106.

Muto, S., Nakayama, M., Potters, J. and Tijs, S.: 1988, On big boss games, *The Economic Studies Quarterly* **39**, 303–321.

Otten, G., Peleg, B., Borm, P. and Tijs, S.: 1998, The mc-value for monotonic ntu-games, *International Journal of Game Theory* **27**, 37–47.

Owen, G.: 1975, On the core of linear production games, *Mathematical Programming* **9**, 358–370.

Potters, J. and Tijs, S.: 1992, The nucleolus of matrix games and other nucleoli, *Mathematics of Operations Research* **17**, 164–174.

Sandsmark, M.: 1999, Production games under uncertainty, *Technical report*, University of Bergen.

Scarf, H.: 1967, The core of an n-person game, *Econometrica* **35**, 50–69.

Schmeidler, D.: 1969, The nucleolus of a characteristic function game, *SIAM Journal of Applied Mathematics* **17**, 1163–1170.

Shapley, L.: 1953, A value for n-person games, *in* A. Tucker and H. Kuhn (eds), *Contributions for the Theory of Games*, Vol. 2, Princeton University Press, Princeton.

Shapley, L.: 1967, On balanced sets and cores, *Naval Research Logistics Quarterly* **14**, 453–460.

Shapley, L.: 1969, Utility comparisons and the theory of games, *La Decision, Aggregation et Dynamique des Orders de Preference*, Edition du Centre National de la Recherche Scientifique, Paris.

Shapley, L.: 1971, Cores of convex games, *International Journal of Game Theory* **1**, 11–26.

Shapley, L. and Shubik, M.: 1969, On market games, *Journal of Economic Theory* **1**, 9–25.

Sharkey, W.: 1981, Convex games without side payments, *International Journal of Game Theory* **10**, 101–106.

Suijs, J. and Borm, P.: 1999, Stochastic cooperative games: superadditivity, convexity, and certainty equivalents, *Games and Economic Behavior* **27**, 331–345.

Suijs, J., Borm, P., De Waegenaere, A. and Tijs, S.: 1999, Cooperative games with stochastic payoffs, *European Journal of Operational Reseacrh* **113**, 193–205.

Suijs, J., De Waegenaere, A. and Borm, P.: 1998, Stochastic cooperative games in insurance and reinsurance, *Insurance: Mathematics & Economics* **22**, 209–228.

van den Nouweland, A., Potters, J., Tijs, S. and Zarzuelo, J.: 1995, Cores and related solution concepts for multi-choice games, *ZOR - Mathematical Methods of Operation Research* **41**, 289–311.

Vilkov, V.: 1977, Convex games without side payments, *Vestnik Leningradskiva Universitata* **7**, 21–24. In Russian.

von Neumann, J. and Morgenstern, O.: 1944, *Theory of Games and Economic Behavior*, Princeton University Press, Princeton.

Wilson, R.: 1968, The theory of syndicates, *Econometrica* **36**, 119–132.

Index

THEORY AND DECISION LIBRARY

SERIES C: GAME THEORY, MATHEMATICAL PROGRAMMING AND OPERATIONS RESEARCH

Editor: S.H. Tijs, *University of Tilburg, The Netherlands*

1. B.R. Munier and M.F. Shakun (eds.): *Compromise, Negotiation and Group Decision.* 1988 ISBN 90-277-2625-6
2. R. Selten: *Models of Strategic Rationality.* 1988 ISBN 90-277-2663-9
3. T. Driessen: *Cooperative Games, Solutions and Applications.* 1988 ISBN 90-277-2729-5
4. P.P. Wakker: *Additive Representations of Preferences.* A New Foundation of Decision Analysis. 1989 ISBN 0-7923-0050-5
5. A. Rapoport: *Experimental Studies of Interactive Decisions.* 1990 ISBN 0-7923-0685-6
6. K.G. Ramamurthy: *Coherent Structures and Simple Games.* 1990 ISBN 0-7923-0869-7
7. T.E.S. Raghavan, T.S. Ferguson, T. Parthasarathy and O.J. Vrieze (eds.): *Stochastic Games and Related Topics.* In Honor of Professor L.S. Shapley. 1991 ISBN 0-7923-1016-0
8. J. Abdou and H. Keiding: *Effectivity Functions in Social Choice.* 1991 ISBN 0-7923-1147-7
9. H.J.M. Peters: *Axiomatic Bargaining Game Theory.* 1992 ISBN 0-7923-1873-0
10. D. Butnariu and E.P. Klement: *Triangular Norm-Based Measures and Games with Fuzzy Coalitions.* 1993 ISBN 0-7923-2369-6
11. R.P. Gilles and P.H.M. Ruys: *Imperfections and Behavior in Economic Organization.* 1994 ISBN 0-7923-9460-7
12. R.P. Gilles: *Economic Exchange and Social Organization.* The Edgeworthian Foundations of General Equilibrium Theory. 1996 ISBN 0-7923-4200-3
13. P.J.-J. Herings: *Static and Dynamic Aspects of General Disequilibrium Theory.* 1996 ISBN 0-7923-9813-0
14. F. van Dijk: *Social Ties and Economic Performance.* 1997 ISBN 0-7923-9836-X
15. W. Spanjers: *Hierarchically Structured Economies.* Models with Bilateral Exchange Institutions. 1997 ISBN 0-7923-4398-0
16. I. Curiel: *Cooperative Game Theory and Applications.* Cooperative Games Arising from Combinatorial Optimization Problems. 1997 ISBN 0-7923-4476-6
17. O.I. Larichev and H.M. Moshkovich: *Verbal Decision Analysis for Unstructured Problems.* 1997 ISBN 0-7923-4578-9

18. T. Parthasarathy, B. Dutta, J.A.M. Potters, T.E.S. Raghavan, D. Ray and A. Sen (eds.): *Game Theoretical Applications to Economics and Operations Research*. 1997 ISBN 0-7923-4712-9
19. A.M.A. Van Deemen: *Coalition Formation and Social Choice*. 1997 ISBN 0-7923-4750-1
20. M.O.L. Bacharach, L.-A. Gérard-Varet, P. Mongin and H.S. Shin (eds.): *Epistemic Logic and the Theory of Games and Decisions*. 1997 ISBN 0-7923-4804-4
21. Z. Yang (eds.): *Computing Equilibria and Fixed Points*. 1999 ISBN 0-7923-8395-8
22. G. Owen: *Discrete Mathematics and Game Theory*. 1999 ISBN 0-7923-8511-X
23. I. Garcia-Jurado, F. Patrone and S. Tijs (eds.): *Game Practice*. 1999 ISBN 0-7923-8661-2
24. J. Suijs: *Cooperative Decision-Making under Risk*. 1999 ISBN 0-7923-8660-4